Student's Guide to the IET Wiring Regulations

Updated to BS 7671:2018+A2:2022
IET Wiring Regulations

Published by The Institution of Engineering and Technology, London, United Kingdom

The Institution of Engineering and Technology is registered as a Charity in England & Wales (no. 211014) and Scotland (no. SC038698).

The Institution of Engineering and Technology is the institution formed by the joining together of the IEE (The Institution of Electrical Engineers) and the IIE (The Institution of Incorporated Engineers).

© 2016, 2018, 2022 The Institution of Engineering and Technology

First published 2015 (978-1-78561-078-3)
Second edition 2018 (978-1-78561-474-3)
Third edition 2022 (978-1-83953-260-3)

This publication is copyright under the Berne Convention and the Universal Copyright Convention. All rights reserved. Apart from any fair dealing for the purposes of research or private study, or criticism or review, as permitted under the Copyright, Designs and Patents Act, 1988, this publication may be reproduced, stored or transmitted, in any form or by any means, only with the prior permission in writing of the publishers, or in the case of reprographic reproduction in accordance with the terms of licences issued by the Copyright Licensing Agency. Enquiries concerning reproduction outside those terms should be sent to the publishers at The Institution of Engineering and Technology, Michael Faraday House, Six Hills Way, Stevenage, SG1 2AY, United Kingdom.

Copies of this publication may be obtained from:
The Institution of Engineering and Technology
PO Box 96, Stevenage, SG1 2SD, UK
Tel: +44 (0)1438 767328
Email: sales@theiet.org
https://electrical.theiet.org

While the author, publisher and contributors believe the information and guidance given in this work are correct, all parties must rely upon their own skill and judgement when making use of them. The author, publisher and contributors do not assume any liability to anyone for any loss or damage caused by any error or omission in the work, whether such an error or omission is the result of negligence or any other cause. Where reference is made to legislation it is not to be considered as legal advice. Any and all such liability is disclaimed.

Permission to reproduce extracts from British Standards is granted by BSI. No other use of this material is permitted. British Standards can be obtained in PDF or hard copy formats from the BSI online shop: https://shop.bsigroup.com/.

ISBN 978-1-83953-260-3 (wiro bound)
ISBN 978-1-83953-262-7 (vitalsource)

Typeset in the UK by The Institution of Engineering and Technology

Printed in the UK by Greens Ltd, Lincoln Road, Cressex Business Park, High Wycombe, Buckinghamshire HP12 3RQ

Contents

List of Figures	7
List of Tables	11
Acknowledgements	13
Welcome	15
What is the right route for you?	15
What will you get from this Guide?	20
The road ahead	20
Section 1 The IET Wiring Regulations	23
1.1 The IET Wiring Regulations	23
1.2 Who is responsible for writing the IET Wiring Regulations?	24
1.3 IET standards and compliance staff	26
1.4 Symbols and SI units	33
1.5 Vocabulary: terms, acronyms and abbreviations	34
1.6 Policies and legal requirements	45
1.7 Guidance on predetermined values in the IET Wiring Regulations	46
Section 2 Health and Safety	51
2.1 Introduction	51
2.2 Risk assessments and method statements	52
2.3 Safe isolation procedure	60
2.4 Tool safety	63

Section 3	**Generation and Transmission**	**71**
3.1	What is used to generate electricity?	71
3.2	What are the characteristics of the electricity generated in the UK?	73
3.3	How is electricity transmitted?	74
3.4	How do we measure what we use?	77
Section 4	**Supply**	**81**
4.1	Supply intake arrangements	81
4.2	Live conductors	81
4.3	Current	82
4.4	Earthing arrangement	84
4.5	Responsibility	87
4.6	Tails	88
4.7	The consumer unit	89
Section 5	**Protection and Isolation**	**95**
5.1	Protective devices	95
5.2	Selecting protective devices	109
5.3	Selectivity	111
5.4	Integration of devices and components	114
5.5	Isolation and switching (Chapter 53 of the IET Wiring Regulations)	114
5.6	Isolators and switches (Chapter 46 of the IET Wiring Regulations)	115
5.7	Isolation	116
5.8	Switching off for mechanical maintenance	118
5.9	Emergency switching off	119
5.10	Functional switching	119
5.11	Undervoltage protection	120
Section 6	**Earthing and Bonding**	**123**
6.1	What is earthing and why do we need it?	123
6.2	Protective equipotential bonding	130
6.3	Supplementary equipotential bonding	135

Section 7 Cable Calculations	139
7.1 Why do we need cable calculations?	139
7.2 Current rating of equipment and protective devices and current-carrying capacity of cables	140
7.3 How design current, protective device rating and current-carrying capacity relate to each other	145
7.4 Voltage drop	151

Section 8 Final Circuits	157
8.1 General installation practices	157
8.2 Designing and installing a new circuit	158
8.3 Luminaires and lighting installations	165
8.4 Radial final circuits (power)	177
8.5 Ring final circuits	179
8.6 Spurs	182
8.7 Shower circuit	184
8.8 Separated extra-low voltage systems (SELV)	185
8.9 Protective extra-low voltage systems (PELV)	186
8.10 Functional extra-low voltage systems (FELV)	188

Section 9 Inspection and Testing	191
9.1 Why inspect and test?	191
9.2 Initial inspection	193
9.3 The tests in detail	195
9.4 Dead tests	197
9.5 Live tests	213

Section 10 Fault Finding	219
10.1 What is a fault?	219
10.2 How to diagnose and locate a fault	222

Section 11 Common Calculations	235
11.1 Simple transposition	235
11.2 Triangle method: voltage, current, resistance and power	237
11.3 Resistors	238

Section 12	**Diversity**	**243**
12.1	What is diversity?	243
12.2	Calculating diversity	244
12.3	Diversity for ring final circuits	246
Section 13	**Prosumer's Electrical Installations**	**251**
13.1	Prosumer's installations	251
13.2	Types of prosumer's electrical installations	253
13.3	Operating modes	253
Appendix A	**Special installations or locations**	**257**
Appendix B	**Tables of symbols**	**271**
Appendix C	**Degrees of protection provided by enclosures (IP code)**	**281**

Answers   **283**

Section 1 The IET Wiring Regulations	283
Section 2 Health and Safety	284
Section 3 Generation and Transmission	284
Section 4 Supply	285
Section 5 Protection and Isolation	285
Section 6 Earthing and Bonding	286
Section 7 Cable Calculations	287
Section 8 Final Circuits	289
Section 9 Inspection and Testing	290
Section 10 Fault Finding	290
Section 11 Common Calculations	291
Section 12 Diversity	291
Section 13 Prosumer's Electrical Installations	293

Index   **295**

List of Figures

Figure 1.1	Extract from the *Wiring Rules and Regulations*, published 1882
Figure 1.2	Examples of JPEL/64 Committee membership
Figure 1.3	Contents page of the IET Wiring Regulations
Figure 1.4	Detail of a regulation number
Figure 1.5	Table 4A2: IET Wiring Regulations
Figure 1.6	Tables 4D1A to 4E2A: IET Wiring Regulations
Figure 1.7	Table 4B1: IET Wiring Regulations
Figure 2.1	Example of a risk assessment
Figure 2.2	Risk assessment flow chart
Figure 2.3	Test leads
Figure 2.4	Safe isolation procedure
Figure 3.1	Generation of electricity
Figure 3.2	Transmission from power station to National Grid
Figure 3.3	Examples of transmission towers
Figure 3.4	Example of a distribution transformer
Figure 3.5	Electricity meters
Figure 3.6	Transmission and distribution network
Figure 4.1a	TN-C-S arrangement
Figure 4.1b	TN-S arrangement
Figure 4.1c	TT arrangement
Figure 4.2	Typical intake layouts and responsibilities
Figure 4.3	Distribution and final circuits
Figure 4.4	Six-way consumer unit with eight modules
Figure 4.5	Split-way board with three ways protected by one RCD and three ways protected by another (12 modules in total)

Figure 5.1	Excerpt from the time/current graph as shown in Appendix 3 of BS 7671
Figure 5.2	HRC system E fuse
Figure 5.3	Cartridge fuse
Figure 5.4	Plug-top fuse
Figure 5.5	BS 3036 semi-enclosed fuse
Figure 5.6	Relationship between I_{CS} and I_{CN} for circuit-breakers
Figure 5.7	Internal components within a circuit-breaker
Figure 5.8	Excerpt from time current graphs for circuit-breakers in Appendix 3 of the IET Wiring Regulations
Figure 5.9	Typical RCD arrangement
Figure 5.10	Examples of selectivity
Figure 5.11	A typical motor circuit
Figure 6.1	TN-C-S arrangement
Figure 6.2	TN-S arrangement
Figure 6.3	TT arrangement
Figure 6.4	Installation without protective equipotential bonding
Figure 6.5	Installation with protective equipotential bonding
Figure 6.6	Unbroken bonding conductor
Figure 7.1	Temperature of environment
Figure 7.2	Cables buried in the ground
Figure 7.3	Depth of the cables buried in the ground
Figure 7.4	BS 3036 fuses
Figure 7.5	Cable grouping
Figure 7.6	Cable run through thermal insulation
Figure 7.7	Cable run underground
Figure 7.8	Checking Table 4B1 from Appendix 4 of the IET Wiring Regulations
Figure 8.1	Recommended arrangement for an RCD protecting a domestic installation
Figure 8.2	Required heights for switches, socket-outlets, etc.
Figure 8.3	Permitted cable routes
Figure 8.4	Protective device
Figure 8.5	Terminals at the ceiling rose
Figure 8.6	Two-plate method

Figure 8.7	Three-plate method
Figure 8.8	Looping power
Figure 8.9	One-way light switch
Figure 8.10	Connection of conductors in a one-way light switch
Figure 8.11	Two-way light switch
Figure 8.12	Two-way conventional
Figure 8.13	Two-way conversion
Figure 8.14	Intermediate light switch
Figure 8.15	Intermediate (conventional) wiring configuration
Figure 8.16	Intermediate (conversion) wiring configuration
Figure 8.17	Even distribution of load over a ring final circuit
Figure 8.18	Taken from Table H2.1 in the *On-Site Guide*
Figure 8.19	Final ring circuit with spurs
Figure 8.20	Connection of cpc to a socket-outlet including earth tail
Figure 8.21	SELV transformer
Figure 8.22	PELV transformer
Figure 8.23	FELV transformer
Figure 9.1	Nulling/zeroing the tester
Figure 9.2	Example of Method 1 on a radial final circuit
Figure 9.3	Example of Method 2 on a radial circuit
Figure 9.4	Typical socket-outlet test adapter
Figure 9.5	Connections for ring final circuit continuity testing: step 1
Figure 9.6	Line from leg one connected to neutral from leg two
Figure 9.7	Line from leg one connected to cpc from leg two
Figure 9.8	Testing insulation resistance between line and neutral
Figure 9.9	Testing insulation resistance between neutral and earth
Figure 9.10	Testing insulation resistance between line and earth
Figure 9.11	Testing polarity using Test Method 1
Figure 9.12	EFLI two-lead test
Figure 9.13	Typical readings displayed on a phase sequence tester
Figure 10.1	Incorrect size terminations or connector blocks
Figure 10.2	Nail through cable
Figure 10.3	Line conductor damaged
Figure 10.4	Poor termination, frayed strands of conductor

Figure 10.5	Conductors incorrectly terminated
Figure 10.6	Ring final circuit
Figure 10.7	Ring final circuit, short circuit between line and neutral
Figure 10.8	Ring final circuit split into two parts
Figure 10.9	Ring final circuit split again
Figure 12.1	Extract from Table A1 *On-Site Guide*
Figure 12.2	Process sequence for calculating diversity for a cooker circuit
Figure 12.3	Ring final circuit
Figure 12.4	Table H2.1 from the IET's *On-Site Guide*
Figure 12.5	Diversity used in different circuits
Figure 13.1	Example of prosumer's electrical installations
Figure C.1	IP code format

List of Tables

Table 1.1	The Wiring Regulations standard-setters
Table 1.2	Types of cable
Table 1.3	Institutions and organizations
Table 1.4	Documents published by the Health and Safety Executive
Table 1.5	Materials
Table 1.6	Expressions you might hear on site
Table 2.1	An everyday risk assessment: crossing the road
Table 2.2	Acceptable risk
Table 2.3	Various hand tools used by electricians
Table 2.4	Various power tools used by electricians
Table 3.1	Generation sources
Table 4.1	Live conductor arrangements within the UK
Table 5.1	Protective devices and relevant Standards
Table 5.2	Protective devices
Table 5.3	Types of installation
Table 9.1	Tests for initial verification
Table 9.2	Minimum values of insulation resistance
Table B.1	General symbols
Table B.2	International system of units: base units
Table B.3	Multiples and sub-multiples of quantities
Table B.4	SI derived units
Table B.5	Symbols for use in schematic wiring diagrams
Table B.6	Making and breaking current
Table B.7	Isolating

Table B.8	Making, breaking and isolating
Table B.9	Meters
Table B.10	Location symbols for installations

Acknowledgements

The Institution of Engineering and Technology acknowledges the invaluable contribution made by the following individuals in the preparation of the Student's Guide to the IET Wiring Regulations.

Institution of Engineering and Technology

S. Devine IEng MIET

We would like to thank the following organizations for their continued support:

Certsure trading as NICEIC

EAL

ECA

Electrical Contractors' Association of Scotland (SELECT)

City & Guilds

Health and Safety Executive (HSE)

NAPIT

We would like to thank the following organizations for their significant contribution to the images used in this Guide:

British Approvals Service for Cables (BASEC)

DeWalt

Socket & See

Stanley Tools

Illustrations by:

Farquhar design: https://farquhardesign.co.uk/ and
G Kenyon Technology Ltd

Cover design and illustration were created by Ken Dobson at Studio Stunt Double: http://studiostuntdouble.com/

Revised, compiled and edited

M. Doughton IEng MIET LCGI

Welcome

If you're reading this, you've most likely decided that a career in the electrotechnical industry is for you.

Whether you're studying a full-time course or you're employed as an apprentice, you've just taken your first step towards a wonderful career in the electrotechnical industry. Electricians are always in high demand, for everything from small household jobs, such as installing light fittings and changing of socket-outlets, through to commercial and industrial electrical installations. It all starts here and now.

Once you've completed an apprenticeship, or achieved the equivalent qualifications and experience, you will be regarded as an electrician. Various organizations provide guidance on what category of electrician you will be: this will depend primarily on your experience and qualifications.

What is the right route for you?

Below, we have listed some of the various steps that can be taken to help you get into the electrotechnical industry and the possibilities that lie ahead.

(a) Full-time courses

Since the introduction of government funding for full-time electrical courses up to Level 3, there has been an increased number of courses available to school leavers. There are also basic courses available for those students who need an introduction into the construction industry. Those who have passed the required Maths and English assessments can enrol immediately onto a Level 2 course. On successful completion of a Level 2 course, students will have the opportunity to progress onto Level 3. At this stage, work experience with a practising electrician is essential in order to obtain the necessary practical experience.

(b) Apprenticeships

NOTE: some employers require applicants to successfully complete a colour deficiency test as well as Maths and English assessments, prior to commencement of employment.

The industry has its ups and downs and there are times when apprenticeships are hard to come by. Luckily, now and again, the government offers financial incentives to employers when they employ an apprentice. Learning providers and local colleges can provide more information on funding and government incentives. It is important to be aware of these incentives when applying to companies for a job, because they do change and not all employers know about them.

Prior to enrolling on an apprenticeship, you will need to be aware of the entry requirements. At the time of publication of this Guide, it is expected that applicants will have at least a Grade 4 (C) GCSE or equivalent in Maths and English. Some colleges will allow their learners to obtain these qualifications during their apprenticeship. This may be quite challenging for some learners, as the Maths and English qualifications will be in addition to the electrical apprenticeship. It is common for colleges and training providers to assess applicants so that they can advise on whether or not they are ready to join the electrical course.

Apprenticeships are slightly different depending on the type of work you do, your employer and, most importantly, where you are. In England, Wales and Northern Ireland, an apprentice will be working towards a Level 3 Diploma. This will usually involve attending college one day a week during term time and working with an electrician for the rest of week. Depending on your employer's preference, block release courses for apprenticeships are also available, where you would attend college for one or two weeks at a time and then return to working with your employer for a similar period. In Scotland, you will attend college on 'block release', which means that you will be at college for a week or more at a time to achieve a Scottish Vocational Qualification (SVQ) in Electrical Installation at SCQF level 7.

Regardless of where you are studying, you will be expected to successfully complete an assessment of competence on completion of your apprenticeship. In England, Wales and Northern Ireland, the assessment of competence is called the Achievement Measurement 2 (AM2 and AM2S). In Scotland, it is called the Final Integrated Competence Assessment (FICA).

The apprenticeship route is ideal if you are a school leaver and have some financial support. But what if you are an adult and looking for a career change? In England and Wales, there are a number of options available, current information on which can be obtained from training centres, colleges and online.

In Scotland, as an adult who is new to the electrotechnical industry, there is only one route to pursue and that is the adult apprenticeship. Applicants will only be considered if they have had at least one year's experience of working in the electrotechnical industry and are employed for the duration of the apprenticeship. Adult apprentices will be expected to attend college on day release (one day every two weeks) and it is expected that they will complete their qualifications within four years of enrolment.

Once you have been accepted onto an apprenticeship, you are well on your way to becoming an electrician. The qualifications you will gain during your apprenticeship will provide you with enough credits to claim what is known as a 'qualification framework'. Once you have this, it can be submitted to various organizations, such as the Joint Industry Board (JIB) or, in Scotland, the Scottish Joint Industry Board (SJIB). You can then apply for your JIB/SJIB grading card, which is recognized throughout Great Britain. See the links below for more information on how to become registered:

www.jib.org.uk
www.sjib.org.uk

(c) Inspection and testing

You have now either completed an apprenticeship or gained the necessary qualifications to be regarded as an electrician. What now?

It doesn't stop there: at this stage, you have just passed your first milestone in your career progression and should now be aiming to achieve a qualification in inspection and testing. Most modern apprenticeships incorporate a mechanism so that learners will automatically gain a qualification for inspection and testing. However, if you are currently enrolled on an apprenticeship, it may be worth checking that this applies to your course. If you have taken an alternative route, such as what is sometimes referred to in the industry as the 'improver route', you may find that you will need to complete this qualification separately.

By now, your understanding of installing wiring systems should be sufficient to allow you to work unsupervised and carry out most types of electrical installation work with confidence.

You can then move on to improve your ability to inspect work carried out by other electricians, as well as to assess existing installations that may be in need of maintenance or upgrading. There are inspection and testing courses available from various awarding organizations. Most people stick with one particular awarding organization, although this is not required. Every qualification must meet a required national standard, so that they are of the same value, regardless of the awarding organization. This essentially means that a Level 3 electrical qualification from one awarding organization is of the same value as a Level 3 electrical qualification from any other awarding organization.

Qualifications for inspection and testing will provide more career opportunities and introduce electricians to a supervisory/ assessment role.

(d) Design

Electricians will, in many cases, be working to, and installing, someone else's design. Designers are commonly electrical engineers who have specifically studied to become an electrical installation designer. However, electricians who have experience as installers are sought after in the design sector, due to their practical understanding of electrical systems. A move into this area may involve more studying and obtaining qualifications in electrical design.

(e) Higher education

It is not uncommon for electricians to continue their education, gaining qualifications such as Higher National Certificates and Diplomas (HNCs and HNDs) and then continuing to achieve a degree in electrical engineering or similar fields.

(f) What next? The world is your oyster...

By the time you reach this stage of your career in the electrotechnical industry, you will most likely be aware of the opportunities that are available to you.

To put it simply...there are lots!

The electrotechnical industry is vast – and growing faster than ever. Technology is fascinating, with the introduction of electric vehicles, wind generation and many other ways of harvesting our renewable energy sources, as well as the development of smart home installations, smart cities, etc. It's a fantastic time for the electrotechnical industry, in the forefront of technological advancement.

(g) Around the world

It is important to remember that a British qualification is recognized in many countries around the world. Popular destinations for qualified electricians are places such as Australia, New Zealand and Canada. Although they have slightly different standards, the principles remain the same.

What will you get from this Guide?

This Guide will provide you with information and references for various aspects of your qualification, as well as the techniques that you will require for a successful career in the electrotechnical industry. Having this Guide by your side will enable you to identify terminologies quickly and see what sections of the IET Wiring Regulations (BS 7671) you need to look through in order to extract the information you need.

We've written this Guide with the aim of helping you understand the IET Wiring Regulations from a student's perspective. The Guide offers clear explanations and examples of how the IET Wiring Regulations form an integral part of the electrotechnical industry and why they should be complied with.

Over and above this, this Guide provides guidance on some situations electricians may come across in their work activities, such as fault finding, risk assessments and legal requirements.

Using this Guide alongside the IET Wiring Regulations and the IET's *On-Site Guide* (a book that provides guidance on the IET Wiring Regulations for installers) will provide the guidance needed while studying for a career in the electrotechnical industry.

The road ahead

Now that you know what possible career opportunities lie ahead, why not make a plan? It is never too early to start!

Career Plan	
Stage 1	For example, student/apprentice: gain experience and knowledge in the electrotechnical industry. Work with electricians to develop skills and understanding.
Stage 2	Achieve qualifications and relevant vocational/practical skills to become an electrician.
Stage 3	Gain additional experience in the industry, offer guidance to apprentices and develop supervisory skills. Gain additional qualifications for inspection and testing and initial verification.
Stage 4	Take on supervisory roles and begin to develop management and design skills; gain qualifications for electrical design, possibly an HNC.
Stage 5	Look at opportunities elsewhere, for example, other companies, other parts of the country and abroad.
Stage 6	Take on managerial roles, design installations, coordinate projects.
Stage 7	Continue to develop qualifications through continuing professional development (CPD); investigate new and emerging areas and technologies.

The IET Wiring Regulations

This Section provides information on the following topics:
- ▶ Who is responsible for the IET Wiring Regulations?
- ▶ Information and guidance relating to the IET Wiring Regulations
- ▶ Finding your way around the IET Wiring Regulations
- ▶ Terminologies, abbreviations and acronyms
- ▶ Guidance on legal requirements
- ▶ Symbols and SI units
- ▶ Guidance on predetermined values in the IET Wiring Regulations

This Section provides information to students to enhance their understanding of the IET Wiring Regulations and identifies the relevance of various qualifications in the electrotechnical industry.

1.1 The IET Wiring Regulations

1.1.1 What are the IET Wiring Regulations and where do they come from?

In 1882 *The Wiring Rules and Regulations* were first issued by the Society of Telegraph Engineers and of Electricians. They consisted of four pages and 21 regulations.

▼ **Figure 1.1** Extract from the *Wiring Rules and Regulations*, published 1882

> **RULES AND REGULATIONS, Etc.** 841
>
> **Society of Telegraph Engineers and of Electricians.**
>
> **RULES AND REGULATIONS**
> FOR THE PREVENTION OF FIRE RISKS ARISING FROM ELECTRIC LIGHTING.
>
> Recommended by the Council in accordance with the Report of the Committee appointed by them on May 11, 1882, to consider the subject.
>
> **MEMBERS OF THE COMMITTEE.**
>
> Professor W. G. Adams, F.R.S., *Vice-President.*
> Sir Charles T. Bright.
> T. Russell Crompton.
> R. E. Crompton.
> W. Crookes, F.R.S.
> Warren De la Rue, D.C.L., F.R.S.
> Professor G. C. Foster, F.R.S., *Past President.*
> Edward Graves.
> J. E. H. Gordon.
> Dr. J. Hopkinson, F.R.S.
>
> Professor D. E. Hughes, F.R.S., *Vice-President.*
> W. H. Preece, F.R.S., *Past President.*
> Alexander Siemens.
> C. E. Spagnoletti, *Vice-President.*
> James N. Shoolbred.
> Augustus Stroh.
> Sir William Thomson, F.R.S., *Past President.*
> Lieut.-Colonel C. E. Webber, R.E., *President.*
>
> These rules and regulations are drawn up not only for the guidance and instruction of those who have electric lighting apparatus installed on their premises, but for the reduction to a minimum of those risks of fire which are inherent to every system of artificial illumination.
>
> The chief dangers of every new application of electricity arise mainly from ignorance and inexperience on the part of those who supply and fit up the requisite plant.
>
> The difficulties that beset the electrical engineer are chiefly internal and invisible, and they can only be effectually guarded against by "testing," or probing with electric currents. They depend chiefly on leakage, undue resistance in the conductor, and bad joints, which lead to waste of energy and the production of heat. These defects can only be detected by measuring, by means of special apparatus, the currents that are either ordinarily or for

1.2 Who is responsible for writing the IET Wiring Regulations?

BS 7671 *Requirements for Electrical Installations* or, as most people call it, the IET Wiring Regulations, is a collection of regulations that are mostly made up of requirements from European Standards, themselves based on international or world standards. There are slight differences in each of these standards due to varying conditions of electrical supply and installation characteristics in different countries. Table 1.1 shows which organization is responsible for each set of standards.

▼ **Table 1.1** The Wiring Regulations standard-setters

 	The world electrotechnical standard-setting body is the International Electrotechnical Commission (the **IEC**). The Standard is IEC 60364 *Low voltage electrical installations*. Any Standards introduced by IEC must usually be adopted by CENELEC and subsequently JPEL/64 within 3 years of being published.
	The European Standards body is the European Committee for Electrotechnical Standardization (**CENELEC**). The Standard is HD 60364 *Low-voltage electrical installations*.
	The British standard-setting body is the British Standards Institution (**BSI**). The standard is BS 7671 *Requirements for Electrical Installations*, more commonly known as the IET Wiring Regulations. You will see on your cover of the IET Wiring Regulations that BS 7671 is followed by a number (for example, ':2018'). This refers to the edition of the Wiring Regulations (so, for example, the 18th Edition, published in 2018).

1.3 IET standards and compliance staff

In the IET Wiring Regulations, on the page titled 'Publication Information', you will see a list of names under the heading **IET Standards & Compliance staff**. These are the people who work directly for the IET and who manage the JPEL/64 Committee and subcommittees.

The JPEL/64 Committee

J stands for Joint (the Committee is jointly run by the BSI and the IET).

P stands for Power (as in 'power generation').

EL stands for Electrical.

64 is the number allocated by the IEC to electrotechnical committees.

The responsibility of the Committee Manager

The Committee Manager has a key role when it comes to the work of the Committee. Before the meeting, the Committee Manager must:

(a) arrange a venue in which Committee members and electrotechnical industry experts can meet;
(b) communicate with industry experts and invite new members onto the Committee;
(c) circulate agendas and the relevant documents to the Committee members – if, for example, there are new electrical products available on the market that may have an impact on the electrotechnical industry, these should be considered by the Committee experts; and
(d) coordinate the documents sent to the relevant Committee members.

During the meeting, the Committee Manager will make a record of everything that is discussed, ensuring that discussions are recorded accurately and any comments made by Committee members are logged. It is important that the Committee Manager can make available upon request any documents that are pertinent to the discussion.

After the meeting, the Committee Manager is required to prepare and circulate details of the discussions in the form of minutes to all Committee members. If there are any decisions that need to be communicated with other institutions or organizations, the Committee Manager has the responsibility of ensuring that the information is passed on. In addition, the Committee Manager will have a technical understanding of the new topics and a knowledge of the technical discussions and agreements made at previous meetings.

Key organizations on the JPEL/64 Committee

A few pages further on in BS 7671, you will see the heading **Joint IET/BSI Technical Committee JPEL/64 Constitution**. Beneath this is a list of all the JPEL/64 Committee members and the organization(s) they represent.

Figure 1.2 shows some of the key organizations represented at the JPEL/64 Committee.

▼ **Figure 1.2** Examples of JPEL/64 Committee membership

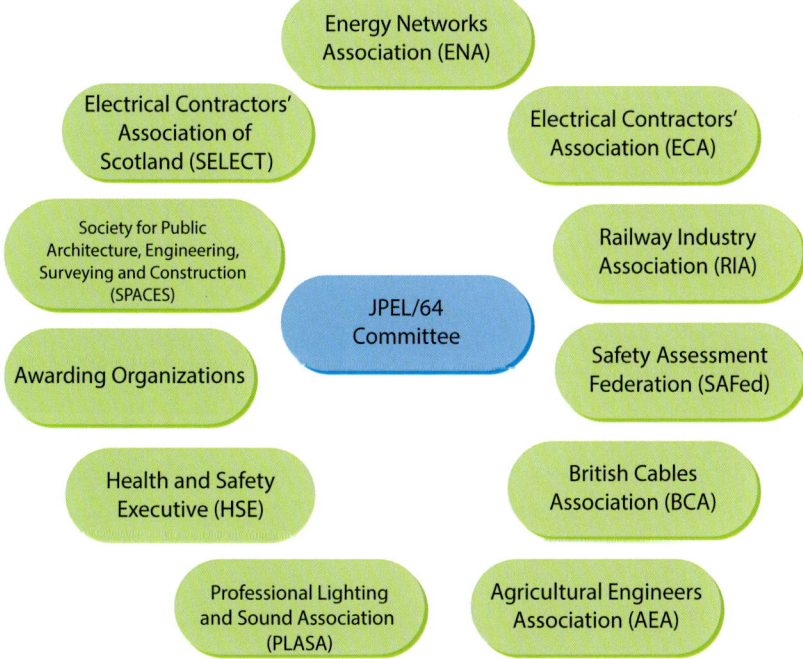

1.3.1 Who are the Committee members and what do they do?

Committee members are carefully selected based on their area of expertise and the organization that they represent. It is important that every point of view is considered before introducing a new regulation or modifying an existing one.

Many of the people responsible for producing the IET Wiring Regulations have worked in the electrotechnical industry and have exceptional knowledge of electrical principles and wiring systems. These include representatives from the Health and Safety Executive (HSE), manufacturers of electrical equipment, experts from the competent person scheme providers and many more.

Most JPEL/64 members are sponsored and supported by their employer or trade organization and have a passion not only for ensuring that the Wiring Regulations for electrical installations meet the needs of users of electrical installations, but also for supporting electrical contractors to maintain practical applications of wiring systems.

1.3.2 Information and guidance for qualifications

The IET Wiring Regulations are regularly updated to meet the demands of ever-improving technology, the need to use our energy as efficiently as possible and, most importantly, to protect persons and livestock from the effects of electric shock.

There are new electrical products entering the market all the time. If there are no standards set for these products, they may perform poorly and, potentially, present a risk to the safety of the user. This also applies to the wiring systems that are designed to supply these products.

While studying for any qualification in the electrotechnical industry, part of the curriculum will involve the use of the IET Wiring Regulations. Studies that will reference the IET Wiring Regulations might range from basic domestic installer courses through to an Honours degree in electrical engineering. More recently, most construction qualifications, such as those on plumbing and heating and ventilation, also reference the IET Wiring Regulations.

While studying to become an electrician, students are required to have a good understanding of the IET Wiring Regulations and how to put them into practice. Regular assessments are carried out throughout courses to confirm that learners are able to understand and interpret the IET Wiring Regulations. When it comes to exam time, it is essential that learners have a copy of the IET Wiring Regulations and the *On-Site Guide*, as these are usually permitted for use during examination by the awarding bodies.

1.3.3 Finding your way around the IET Wiring Regulations

When you first open the IET Wiring Regulations, you're bombarded with numbers and text. It can be difficult to make any sense of these and even more difficult to source the information that is needed. In this Section, the basic structure of the content is explained.

Parts and Appendices

The Regulations are divided into 8 Parts and 17 Appendices two of which are not used. Each Part is made up of Chapters and Sections, which are themselves made up of the Regulations. The Appendices, which are included at the back of the book, are used to provide additional detail relating to the Regulations. Details of these Parts and their Chapters can be found in the contents page at the front of the IET Wiring Regulations (see Figure 1.3).

▼ **Figure 1.3** Contents page of the IET Wiring Regulations

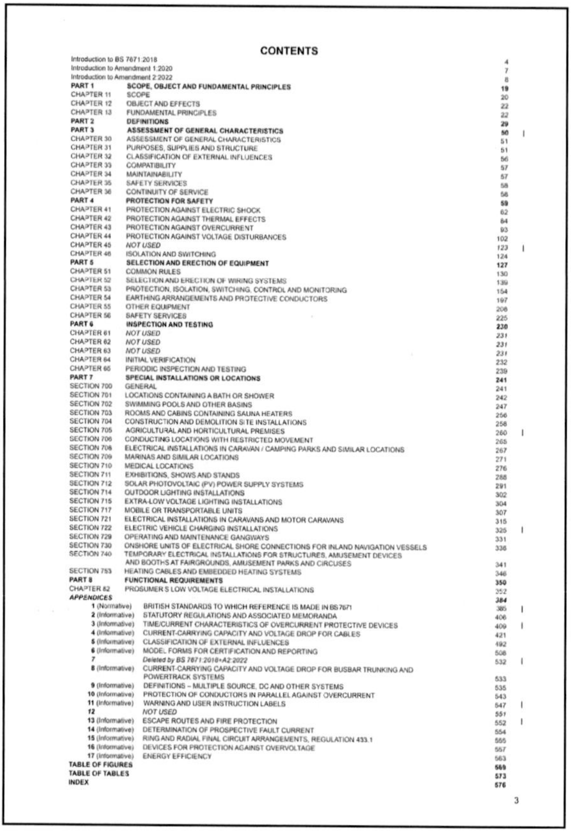

You can familiarize yourself with the Parts to begin with:

Parts

1. Scope, object and fundamental principles
2. Definitions
3. Assessment of general characteristics
4. Protection for safety
5. Selection and erection of equipment
6. Inspection and testing
7. Special installations or locations (this part of the IET Wiring Regulations has been divided into Sections, rather than Chapters)
8. Functional requirements

Next, you can look at the Appendices:

Appendices

1. British Standards to which reference is made
2. Statutory regulations and associated memoranda
3. Time/current characteristics of overcurrent protective devices and RCDs
4. Current-carrying capacity and voltage drop for cables
5. Classification of external influences
6. Model forms for certification and reporting
7. Not used
8. Current-carrying capacity and voltage drop for busbar trunking and powertrack systems
9. Definitions – multiple source, DC and other systems
10. Protection of conductors in parallel against overcurrent
11. Warning and user instruction labels
12. Not used
13. Escape routes and fire protection
14. Determination of prospective fault current
15. Ring and radial final circuit arrangements, Regulation 433.1
16. Devices for protection against overvoltage
17. Energy efficiency

Regulations

Now we can look at an individual regulation. The regulation number below has been broken down into its relevant categories (Figure 1.4).

Figure 1.4 Detail of a regulation number

514.5.2 Protective condutor

514.5.2

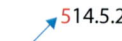

5 indicates the part number. Selection and Erection.

514.5.2

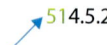

51 indicates the chapter number. Common rules.

514.5.2

514 indicates the section. Identification and notices.

514.5.2

5 indicates the group. Indentification of conductors by letters and/or numbers.

514.5.2

2 indicates the regulation number. Protective conductor Conductors with green-and-yellow colour identification shall not be numbered other than for the purpose of circuit identification.

You will notice many regulations that end with a 200 number. For example 201, 202, 203 and so on. When a regulation ends with a 200 number it means that it is a UK-specific regulation and has not been adopted from a European standard.

Test your knowledge
Have a look at the following regulation and list the part, chapter, section and regulation:
462.1.201
Part:
Chapter:
Section:
Regulation:

1.4 Symbols and SI units

Appendix B of this Guide sets out symbols and the international system of units (SI units). It is important to be familiar with these symbols and units when preparing to work in the electrotechnical industry. It is also important that drawings are easy to understand, and this includes the symbols that are used. Where no standard symbol is available, a specialist symbol can be used.

When symbols are used in technical drawings, they should be explained in a key or legend. In particular, a more detailed legend, specifying areas such as lighting design, the type of fitting, manufacturer source output and the catalogue reference number of products might be required to help the reader understand.

The symbols and SI units included in Appendix B can also be found in the IET's *On-Site Guide*. They are extracted mainly from IEC 60617 *Graphical symbols for diagrams*, supplemented by references from other standards.

1.5 Vocabulary: terms, acronyms and abbreviations

Various descriptions and names are shortened in the IET Wiring Regulations, to speed up the process of communication. If everything was spoken or written in full, it would take considerably longer to get information across. Remember:

(a) terminology can vary internationally and even from town to town. However, when it comes to engineering terminology, it is, in the most part, standardized across the UK.
(b) abbreviations are generally shortened words and, in some cases, are a combination and shortening of two or more words.
(c) acronyms are the abbreviation formed from the initial letters of other words and pronounced as a word.

Whether studying electrical qualifications or working in the electrotechnical industry, you will come across a great number of phrases, terminologies, acronyms and abbreviations.

It's important to familiarize yourself with as many of these as possible. Doing so will provide you with the ability to interpret the IET Wiring Regulations and the various related publications quickly and easily.

1.5.1 Common terms

While studying for an electrotechnical qualification, Table 1.2 will be useful for reference.

▼ **Table 1.2** Types of cable (Images reproduced with permission by BASEC)

ADSL	ADSL, or **asymmetric digital subscriber line**, is generally used for transmitting data.
CAT 5,6,7...	Category 5,6,7: generally used for computer networks such as Ethernet and data transfer.
Conductor	The essential part of the cable that carries electrical current. It is generally made of copper, but can also be made of aluminium, gold and various other metals, including alloys.
cpc	A cpc, or **circuit protective conductor**, can be a conductor or cable and is designed to carry current in the event of a fault, i.e. line to exposed- or extraneous-conductive-part.
Flex (flexible)	Flexible cable, often used for connecting portable appliances.

▼ **Table 1.2** *cont.*

Fire resistant cables to BS 7629	These cables are sometimes considered as an alternative to MIMS. Commonly used for the installation of fire and smoke detection systems, due to its high resistance to heat and fire.
Insulation/insulator	A layer of non-conducting material that surrounds the conductor to prevent electrical current leaking.
MICC MIMS PYRO	MICC is a **mineral-insulated copper-clad cable** and MIMS is a **mineral-insulated metal-sheathed** cable. PYRO is a brand name that is commonly used to refer to MICC and MIMS and is essentially the same thing.
Multi-core	This is a cable that consists of (generally) more than two conductive cores, excluding the cpc.
Sheathing inner/outer	A protective layer that surrounds the insulation. It is designed to offer protection from mechanical damage (knocks and bumps) and is commonly made from PVC.
Single(s)	Cable(s) that consists of one conductor. It is ordinarily made up of twisted strands protected by an insulator.

▼ Table 1.2 cont.

SWA 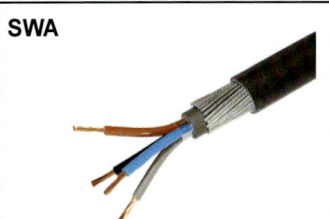	SWA, or **steel wire armour**, is a cable that has additional steel wire strands twisted around the sheathing to provide protection from tougher impacts. The cable is commonly used outside or buried in the ground.
SY YY CY 	This type of cable has a light braided steel armour that provides protection from light mechanical impact. The 'S' refers to steel braid and the 'Y' refers to PVC. YY is usually grey PVC sheath with PVC insulated flexible conductors. CY is tinned copper wire braid, usually grey PVC sheath with PVC insulated flexible conductors. For the purpose of this guidance and for compliance with BS 7671, the use of non-standard cable such as SY, CY and YY is discouraged.
T&E 	T&E, or flat **twin and earth**, PVC flat profile, two-core and earth, is a flat PVC cable that consists of three conductors. Two of the conductors have insulation (brown and blue) and one conductor has no insulation and is designed to be used as a cpc.
Thermoplastic (PVC)	PVC, or **polyvinyl chloride**, is a material that is commonly used to insulate and sheath electrical cable. This cable comes in many different permutations and conductor sizes.
Thermosetting XLPE 	XLPE, or **cross-linked polyethylene**, is a material used to insulate and sheath cable that has a high operating temperature, e.g. 90 °C. This cable comes in many different permutations and conductor sizes.

▼ Table 1.2 *cont.*

Conductive parts	
Exposed-conductive-part	A conductive material that is part of an electrical installation, which is not intended to be live, but which may become live in the event of a fault, for example, a metal case of a light fitting or the metal casing of a kettle.
Extraneous-conductive-part	A conductive material that does not form part of the electrical installation and is liable to introduce a potential, generally Earth potential. This may also become live in the event of a fault, for example, structural steel that is in contact with the general mass of the Earth.

	Extra-low voltage
ELV	ELV, or **extra-low voltage**, is defined in the IET Wiring Regulations as a nominal voltage not exceeding 50 V AC or 120 V ripple-free DC, whether between conductors or to Earth.
FELV 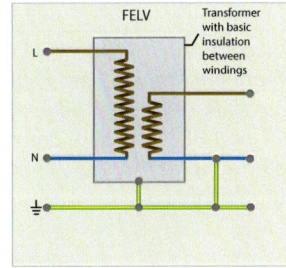	FELV, or **functional extra-low voltage**, is any other ELV circuit that does not meet the requirements for a SELV or PELV circuit. Although the FELV part of a circuit uses an ELV, it is not protected from accidental contact with higher voltages in other parts of the circuit.
PELV 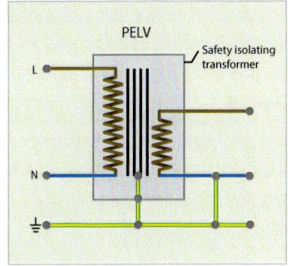	Unlike SELV, a PELV (**protective extra-low voltage**) circuit can have a protective earth. A PELV circuit, as with SELV, should be designed to guarantee a low risk of accidental contact with a higher voltage. For a transformer, this means that the primary and secondary windings must be separated by an extra insulation barrier.
SELV 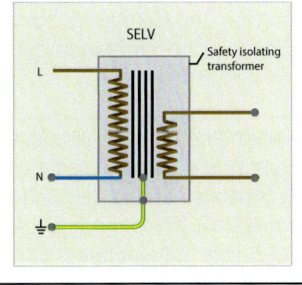	SELV, or **separated extra-low voltage**, is an ELV supply (less than 50 V AC) taken from a safe source, for example, a Class II isolating transformer that complies with BS EN 61558-2-6.

1.5.2 Institutions and organizations

A lot of work goes on behind the scenes in the electrotechnical industry, much of which is achieved by the following institutions and organizations. It's useful to know how these institutions and organizations are referred to.

▼ **Table 1.3** Institutions and organizations

BS 7671	British Standard (BS) 7671 *Requirements for Electrical Installations*. Published jointly by the IET and BSI and more commonly referred to as the IET Wiring Regulations.
BS EN	British Standard and European Normalization (Standardization): an EN that has been implemented in the UK.
BSI	British Standards Institution. BSI is an independent national body responsible for preparing British standards.
CENELEC	In French: Comité Européen de Normalisation Électrotechnique. In English: the European Committee for Electrotechnical Standardization.
HSE	Health and Safety Executive. The Health and Safety Executive (HSE) is Britain's national regulator for workplace health and safety. The HSE acts in the interest of the general public to reduce work-related deaths and serious injury in Great Britain.

▼ Table 1.3 cont.

IEC	International Electrotechnical Commission. The IEC is an international standards and conformity assessment body for all fields of electrotechnology. IEC Standards range from power generation, transmission and distribution to home appliances and office equipment and cover almost anything that forms part of an electrical system.
IET	Institution of Engineering and Technology. The IET was formed by the joining of the IEE (Institution of Electrical Engineers) and the IIE (Institution of Incorporated Engineers).
ISO	International Organization for Standardization. ISO is a voluntary organization whose members are recognized authorities on standards, each one representing one country. ISO develops standards ranging from manufactured products and technology to food safety, agriculture and healthcare.
JPEL/64	The decision-making body for BS 7671. See Section 1.3 for more details.
ECA	The Electrical Contractors' Association (ECA) is the UK's largest trade association representing electrotechnical and engineering services organizations at the regional, national and European level.
SELECT	SELECT is the trade association for the electrical contracting industry in Scotland. It was founded in 1900 as The Electrical Contractors' Association of Scotland.

▼ Table 1.3 *cont.*

Awarding organizations	
C&G	City & Guilds
EAL	Excellence Achievement and Learning
Competent Person Scheme providers	
NAPIT	The National Association of Professional Inspectors and Testers (NAPIT) is one of the leading government-approved and United Kingdom Accreditation Service (UKAS) accredited membership scheme operators in the building services and fabric sector.
NICEIC	The National Inspection Council for Electrical Installation Contracting (NICEIC) is a voluntary regulatory body for the electrical contracting industry. It has been assessing the electrical competence of electricians for over 65 years. It currently maintains a roll of over 26,000 registered contractors and provides certification to firms across the heating, plumbing, renewable and insulation sectors.

1.5.3 Acts and regulations

During your studies or when you're working as an electrician, you might come across these acts and regulations.

▼ **Table 1.4** Documents published by the Health and Safety Executive

COSHH 2002	Control of Substances Hazardous to Health. COSHH ensures that adequate training is provided to staff on the safe storage, handling and use of potentially harmful chemicals.
EAWR 1989	Electricity at Work Regulations. The EAWR ensure that working conditions in the workplace do not put employees at risk of electric shock.
ESQCR 2002	Electricity Safety, Quality and Continuity Regulations 2002, as amended. The ESQCR is a legal document known as a Statutory Regulation (but more formally called a 'Statutory Instrument' – its number is SI 2002/2665). It sets out the requirements that suppliers of electricity must follow, ensuring that the electrical supply is safe and adequate for the supply requirements.
HSWA 1974	Health and Safety at Work etc. Act 1974. The HSWA makes further provision for securing the health, safety and welfare of persons at work and for protecting others against risks to health or safety in connection with the activities of persons at work.
MHOR 1992	Manual Handling Operations Regulations. The employer's duty is to avoid manual handling as far as reasonably practicable if there is a possibility of injury.
PPEWR 1992	Personal Protective Equipment at Work Regulations. The responsibility lies with the employer to assess what personal protective equipment (PPE) is required for the job. The responsibility to wear such PPE lies with the employee.
PUWER 1998	Provision and Use of Work Equipment Regulations. PUWER covers the use of tools and equipment used for work activities.

▼ Table 1.4 *cont.*

RIDDOR 2013	Reporting of Injuries, Diseases and Dangerous Occurrences Regulations 2013. RIDDOR puts duties on employers, the self-employed and people in control of work premises (the 'responsible person') to report certain serious workplace accidents, occupational diseases and specified dangerous occurrences (near-misses).

▼ Table 1.5 Materials

Cable ladder (sometimes referred to just as a 'ladder') 	As it sounds, a length of plastic or metal material that looks like a ladder, normally used to provide support for heavy cables. NOTE: Cable tray and ladder systems are not considered to be 'containment', as they don't completely cover or contain the cables.
Cable tray (sometimes referred to just as a 'tray') 	A tray is made up of perforated sheets of metal or plastic with turned-up sides and is commonly used in commercial and industrial environments.
Conduit 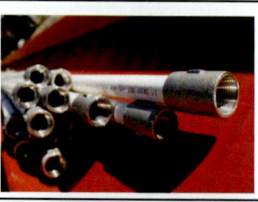	A metal or plastic pipe that commonly ranges from 16 mm to 32 mm in diameter and usually comes in 3-metre lengths. A diameter of up to 63 mm is available for larger installations. Often generally referred to as 'containment'.
Trunking 	Trunking is a metal or plastic U-shaped compartment with a removable lid that is used to contain cables. Trunking can come in a wide variety of sizes and with multi-compartments. Often generally referred to as 'containment'.

▼ **Table 1.6** Expressions you might hear on site

Cable pulling	When installing a cable, installers may work together to pull a very long length of cable into a building to ensure that it is pulled through the building safely without being damaged.
Chasing	Making a channel in a surface, normally brick or plaster, so that a cable can be installed and plastered over.
De-energized	An item of equipment or part of an electrical installation that has been switched off or isolated.
First fix	The first of two stages of installing a wiring system. This stage will involve preparing and installing containment for cable, pulling cables into the containment and fixing or screwing boxes for light switches, socket-outlets, etc. to walls.
Fixing	The action of securing a piece of equipment to a wall or ceiling, for example, securing a conduit saddle to a wall using a screw and wall plug.
Glanding	The action of connecting a cable, such as SWA, to an accessory. Glands are purpose-made for a particular type, size and/or shape of cable and are not usually interchangeable.
Reaming	When a hole or length of conduit has been cut, the excess sharp burrs or pieces of metal or any other remnant material must be removed. This process is called 'reaming' and the tool is called a 'reamer'.
Second fix	The second of two stages of installing a wiring system. This stage will involve terminating electrical equipment and accessories and fitting them into their finished location. The second fix is usually carried out in the final stages of construction.
Stripping out	The removal of redundant electrical installations.

1.6 Policies and legal requirements

The IET Wiring Regulations are non-statutory, which means that they are not law. They may, however, be used in a court of law in evidence, to claim compliance with statutory requirements. The relevant statutory provisions are listed in Appendix 2 of the IET Wiring Regulations and include Acts of Parliament and the regulations that those Acts have introduced. In some cases, statutory regulations may be accompanied by codes of practice approved under Section 16 of the Health and Safety at Work etc. Act (HSWA) 1974. In the *Electricity at Work Regulations 1989, Guidance on Regulations* (HSR25), published by the HSE, it states that "BS 7671 is a code of practice which is widely recognized and accepted in the UK and compliance with it is likely to achieve compliance with relevant aspects of the Electricity at Work Regulations 1989".

The ESQCR set out the parameters within which suppliers of electricity have to operate. These include frequency, voltage and provision of an earth.

1.7 Guidance on predetermined values in the IET Wiring Regulations

As you begin your career in the electrotechnical industry, you'll realise that having a good understanding of maths is essential. During your time at college, you will be required to perform calculation after calculation – long equations that don't seem to have any relevance to the work that you're doing. It is, however, important to understand these calculations. Fortunately, the IET and the electrotechnical committees have done a lot of the hard work for you.

When looking through the IET Wiring Regulations, you will encounter lots of predetermined values that may not make a lot of sense: tables full of correction factors, maximum earth fault loop impedance (EFLI) values, voltage drop, grouping factors and many more.

These values have been accurately calculated and made available in the IET Wiring Regulations and related guidance to help electricians and designers in the electrotechnical industry to standardize their work and maintain safe wiring systems and working practices.

In this section, there are some examples to help you understand the importance of accurate values when designing an electrical system.

1.7.1 Installation methods

Table 4A2 from the IET Wiring Regulations shows, with the aid of diagrams, the various installation methods that can be used and the correct values to use when working out the current-carrying capacity of cables that are installed in a certain way.

There are over one hundred examples like the one below. Notice that the reference method for this type of installation is 'B': you'll need to remember this later when doing cable calculations.

▼ **Figure 1.5** Table 4A2: IET Wiring Regulations

Installation Method			Reference Method to be used to determine current-carrying capacity
Number	Examples	Description	
21		Single-core or multicore cables: - fixed directly under a wooden or masonry ceiling	B (Higher than standard ambient temperatures may occur with this installation method)

1.7.2 Current-carrying capacity of conductors

The following example shows the predetermined current-carrying capacity values for cables installed using various methods. To determine these values takes a considerable amount of time and involves a lot of calculation, so would therefore place additional cost on the customer. Factors that need to be considered in determining the values include the following:

 (a) how the resistance of a conductor is affected by temperature;
 (b) the method of installation, such as surface-mounted, trunking etc.;
 (c) whether it is installed through or alongside thermal insulation;
 (d) whether it is a single-phase or three-phase circuit;
 (e) whether it is AC or DC; and
 (f) others.

Such consideration is very difficult and time-consuming and the IET therefore provides tables that include the information required to design an installation.

Tables 4D1A to 4E2A of the IET Wiring Regulations show the current-carrying capacity of various types and sizes of cable using different installation methods.

▼ **Figure 1.6** Tables 4D1A to 4E2A: IET Wiring Regulations

TABLE 4D1A – Single-core 70 °C thermoplastic insulated cables, non-armoured, with or without sheath
(COPPER CONDUCTORS)

CURRENT-CARRYING CAPACITY (amperes):

Ambient temperature: 30 °C
Conductor operating temperature: 70 °C

Conductor cross-sectional area	Reference Method A (enclosed in conduit in thermally insulating wall etc.)		Reference Method B (enclosed in conduit on a wall or in trunking etc.)		Reference Method C (clipped direct)		Reference Method F (in free air or on a perforated cable tray horizontal or vertical)				
							Touching			Spaced by one diameter	
	2 cables, single-phase a.c. or d.c.	3 or 4 cables, three-phase a.c.	2 cables, single-phase a.c. or d.c.	3 or 4 cables, three-phase a.c.	2 cables, single-phase a.c. or d.c. flat and touching	3 or 4 cables, three-phase a.c. flat and touching or trefoil	2 cables, single-phase a.c. or d.c. flat	3 cables, three-phase a.c. trefoil	3 cables, three-phase a.c trefoil	2 cables, single-phase a.c. or d.c. or 3 cables three-phase a.c. flat	
										Horizontal	Vertical
1	2	3	4	5	6	7	8	9	10	11	12
(mm²)	(A)	(A)	(A)	(A)	(A)	(A)	(A)	(A)	(A)	(A)	(A)
1	11	10.5	13.5	12	15.5	14	-	-	-	-	-
1.5	14.5	13.5	17.5	15.5	20	18	-	-	-	-	-

The reason these values change depending on the method of installation is down to the basic principles of electricity. When current passes through a conductor, it generates heat. When conductors get hotter, the resistance of the conducting material, i.e. copper, increases. If that conductor is contained in thermal insulation, it will retain the heat and gradually get hotter and will therefore have less current-carrying capacity than it would if it were hanging free in the air. When looking at Tables 4D1A to 4E2A, you can see that the least enclosed cables have a higher current-carrying capacity than the same cables enclosed in thermal insulation.

1.7.3 Factors

When calculating the correct size of cable to use for a particular installation, it is essential that the tables provided in the IET Wiring Regulations and other relevant guidance are used. These values are factors, most of which are based on the thermal effects of certain types of installation methods.

An example of a table within the IET Wiring Regulations listing factors is shown here.

▼ **Figure 1.7** Table 4B1: IET Wiring Regulations

TABLE 4B1 – Rating factors (C_a) for ambient air temperatures other than 30 °C

Ambient temperature^a °C	Insulation				
	60 °C thermosetting	70 °C thermoplastic	90 °C thermosetting	Mineral^a	
				Thermoplastic covered or bare and exposed to touch 70 °C	Bare and not exposed to touch 105 °C
25	1.04	1.03	1.02	1.07	1.04
30	1.00	1.00	1.00	1.00	1.00
35	0.91	0.94	0.96	0.93	0.96

There are many more tables in the IET Wiring Regulations that provide factors for various different types of equipment and installation methods.

It is important to remember that there are a number of factors that need to be considered for any electrical installation. The IET Wiring Regulations and associated guidance provide electricians with the information they need to design and install safe and efficient wiring systems.

1.7.4 Earth fault loop impedance (EFLI)

Once an installation has been completed, there are various tests that must be carried out before the circuit is energized, to ensure that the installation meets the requirements of the IET Wiring Regulations. The committee of experts (JPEL/64) has calculated the impedance values that need to be achieved to ensure that protective devices disconnect within the required times given in Table 41.1 of the IET Wiring Regulations (more information on this can be found in Section 6 (Earthing and Bonding) and Section 9 (Inspection and Testing) of this Guide).

1

		Test your knowledge
	1.	When were the Wiring Regulations first issued?
	2.	Who is responsible for electrical regulations internationally?
	3.	At what stage of construction would you normally carry out the second fix of an electrical installation?
	4.	In which Appendix of the IET Wiring Regulations will you find descriptions of various installation methods?
	5.	What is the name of the UK committee that is responsible for the IET Wiring Regulations?
	6.	In which Part of the IET Wiring Regulations will you find the additional requirements for Special Installations or Locations?
	7.	What statutory regulation relates to the need for the electrical supply to be safe and consistent throughout the UK?
	8.	Where can you find descriptions and definitions in the IET Wiring Regulations?
	9.	What does PPEWR stand for?
	10.	What set of regulations relate to employees using electrical equipment at work?

Health and Safety 2

This Section provides information on the following topics:
- ▶ Risk assessment processes
- ▶ Safe isolation procedures
- ▶ Tool safety

2.1 Introduction

There is a risk to the health and safety of anybody, anywhere. The Health and Safety at Work etc. Act 1974 (also referred to as HSWA, the HSW Act, the 1974 Act or HASAWA) is a very important piece of legislation. It puts the responsibility on employers to provide, so far as is reasonably practicable, a safe working environment for employees and sufficient training so that they can work safely.

This subject always seems to be the least favoured, along with regulatory requirements. However, the implementation of health and safety in the construction industry has significantly reduced the number of accidents, injuries and fatalities over the years.

Health and safety begins at the design stage and continues right through to the end of any project. A great number of factors must be considered, ranging from the location of the construction project through to the environmental impact of the work being carried out. Individual work activities that will take place must also be considered, no matter how small. If there is any demolition work to be carried out, the existing building fabric must be assessed for any hazardous materials, such as asbestos.

2.2 Risk assessments and method statements

In this Section we will explain what a risk assessment is and list the fundamental requirements that an electrician should consider when designing and testing an electrical installation, installing equipment, and so on.

2.2.1 What is a risk assessment?

Most of us don't realize that we carry out risk assessments every minute of every day. While we are assessing these risks, we are immediately thinking of various steps to take to avoid them. For instance, if we approach a banana skin on the pavement, we will automatically avoid stepping on it, as we know that we could slip on it and injure ourselves; we have, therefore, carried out a risk assessment and implemented a strategy to lower the risk of injury.

▼ Table 2.1 An everyday risk assessment: crossing the road

Risk	Risk reduction
Possible collision with a car.	Look both ways and listen.
May trip while standing down from the kerb.	Check the kerb height before stepping down.
Restricted access on the other side.	Plan where you want to cross and check the other side.

Table 2.1 lists a number of risk factors that we consider could lead to an accident or injury – even though we don't consciously decide to conduct a risk assessment every time we cross the road.

A risk assessment will, in most cases, need to be documented and accompanied by a method statement. If you are working on your own or have less than five employees, it is not a legal requirement to produce a written risk assessment. However, this does not mean that a risk assessment should not be carried out.

2.2.2 What happens when a risk assessment is not carried out?

Example

On a large project in Cambridge, a team installing plasterboard were preparing for the following week's work by loading all their material and tools into the building. This was done on a Friday afternoon. They had decided to stack all the plasterboard sheets in one room on the 4th floor central to the area they would be working in. This meant that there were around two tonnes of plasterboard in a single room in the building, in addition to tools and equipment.

Over the weekend, a security guard was required to inspect the entire construction site periodically to ensure that there were no trespassers or hazardous situations. While doing his last check at around 05:30 on Monday morning, he heard a loud crashing sound close to his location in the building. When he went to inspect, he found himself looking down from the 4th floor to the underground car park 20 metres below. The weight of the plasterboard sheets had been more than the floor was designed to withstand and it had collapsed under the weight, taking out the floor in every room below until it hit the concrete floor at the bottom. Not more than two hours later, up to ten construction workers were due to be working on the floors below. This could have had fatal consequences if the floor had held out for a little while longer.

If a risk assessment had been carried out, it is likely that the weight of the material would have been taken into consideration and the accident would not have happened, preventing the damage caused and the risk of possible injury.

Consequences of not carrying out a risk assessment

Organizations have a legal duty to put in place suitable arrangements to manage health and safety. If an individual or an employer does not carry out a risk assessment (which must be documented, if employing more than five people), there may be legal consequences if an injury occurs as a result of a preventable risk. A risk assessment may have identified the possible risk and injury could have been avoided.

2.2.3 What is a method statement?

A method statement is a document that details how a task is to be completed and should outline the potential risks that may be involved and include guidance on how to do the job safely. The method statement must also include any measures that have been introduced to ensure the safety of anyone who is affected by the work being done.

> **Example of risk assessment considerations**
>
> You have been asked to install additional lighting in a school classroom during school holidays. One of the first things that you must complete as an installer/designer is a risk assessment.
>
> When carrying out a risk assessment for the installation of basic electrical equipment, the installer will consider where and how the equipment needs to be installed:
> - is it easily accessible?
> - will a ladder, steps or tower be required?
> - how heavy is the equipment that must be installed?
> - will more than one person or lifting device be needed to install the equipment?
> - to what support is the equipment being fixed?
> - is the building fabric hazardous in any way (for example, is asbestos present)?
> - what is the building used for?
> - will there be members of the public or employees that will be affected by the work to be carried out?
>
> The more concise a risk assessment, the easier it is to reduce the potential injury or inconvenience that may occur.

We can now look at the various measures that can be implemented to reduce the risks that have been identified. When documenting these measures, you are producing a method statement.

Figure 2.1 is an example of a risk assessment relating to two of the possible risks that may be identified prior to carrying out the small job.

▼ **Figure 2.1** Example of a risk assessment

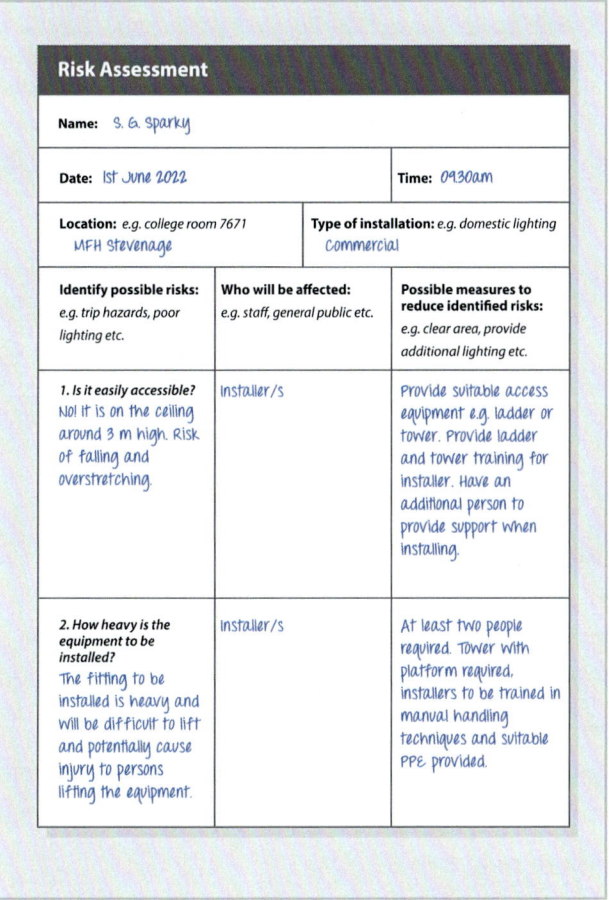

2.2.4 Acceptable risk

If the measures to reduce some of the possible risks completely are expensive and time-consuming, it may be necessary to categorize these as 'acceptable risks' and take reasonable measures to prevent the possibility of accident or injury. You can use a risk rating system like the one below, which will help you to determine the level of severity of the identified risks.

▼ **Table 2.2** Acceptable risk

Risk rating			Risk action bands	
Chances of occurring	X Level of injury	=	Risk assessed as	Measures of control
1. Very unlikely	1. Minor injury		Minimal 1 or 2	Maintain existing measures.
2. Not likely	2. Slight injury		Low 3 or 4	Review measures.
3. Likely	3. Serious injury		Medium 6 or 8	Improve measures.
4. Very likely	4. Major injury		High 9, 12 or 16	Improve measures immediately. May require work to discontinue until additional measures have been applied to reduce the risk.

Minimal risk: 1 or 2

If there are measures in place to maintain a minimum risk, there is no immediate action required. The risk assessment will be continuously reviewed so that any hazards arising will be identified and rectified.

Low risk: 3 or 4

Even a low risk can be reduced in most situations. There may still be something you can do to reduce the rating to minimal.

Medium risk: 6 to 8

A rating of 6 or 8 is quite risky. You should review existing safety measures and apply as many measures as possible to reduce the risk rating to minimal or at least low.

High risk: 9 to 16

This is obviously quite a risky situation. If you cannot identify sufficient measures to reduce the risk further, then it may be necessary to consult with a professional advisor so that adequate provisions can be applied to reduce the risk factor. It may be that additional training or specialist equipment is needed to achieve this.

2.2.5 What next?

The risk assessment doesn't stop there. It is a requirement that the risk assessment is continuously reviewed for the duration of the work being done. This is necessary, as nothing stays the same for long. On a building site, there are changes and developments occurring daily, if not hourly. Some risks may be reduced, while others are increased. It's therefore important to keep the risk assessment up to date.

> **Remember...**
>
> **A summary of the process**
>
> (a) Assess the possible risks – it is advisable to make a record of these.
> (b) Determine the possible consequences of the risks identified – again, details of the possible consequences should be documented.
> (c) Document who or what will be affected.
> (d) Put together a list of measures that can be applied to reduce the impact or prevent the risk altogether.
> (e) Evaluate the risk assessment and add to it where necessary. For example, when materials are moved from one location to another, there may be additional risk at the new location.
> (f) Continuously review the risk assessment and update accordingly.
>
> Following these steps should satisfy most organizations that a sufficient risk assessment has been carried out. For apprentices building a portfolio for a vocational qualification, it is important to document that these steps have been followed.

▼ **Figure 2.2** Risk assessment flow chart

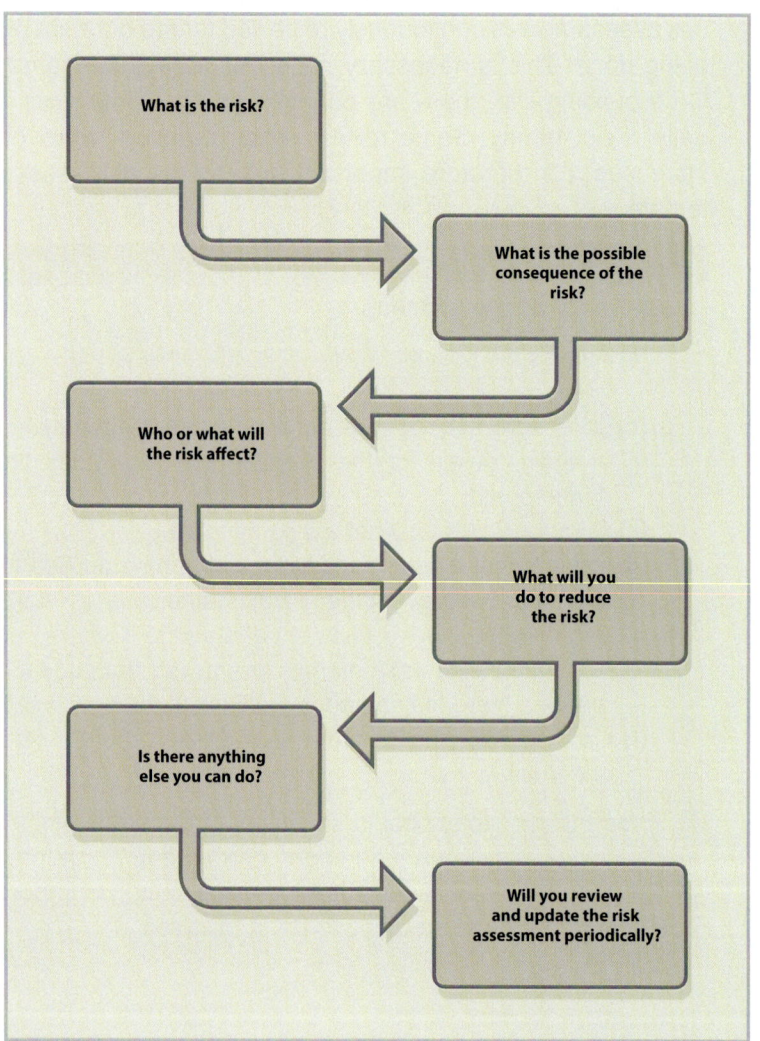

Using the template below, write a risk assessment for a work activity you will be doing in the workshop, such as installing a lighting circuit, cutting conduit, etc.

Remember, you can't remove a risk completely, but there are ways to reduce the risk.

Test your knowledge

Risk Assessment

Name:

Date: | **Time:**

Location: e.g. college room 7671 | **Type of installation:** e.g. domestic lighting

Identify possible risks: e.g. trip hazards, poor lighting etc.	Who will be affected: e.g. staff, general public etc.	Possible measures to reduce identified risks: e.g. clear area, provide additional lighting etc.
1.		
2.		
3.		
4.		
5.		

2.3 Safe isolation procedure

The safe isolation procedure is a process that, when carried out correctly, will reduce the risk of electric shock, burns or death of any person likely to come into direct contact with an electrical installation, whether that installation is in the process of being installed or is under maintenance. It is also very important to inform the person responsible for the premises of the implications of the procedure.

If the procedure has not been carried out correctly, and the protective device has not been locked off and labelled with an approved device, there is the possibility that somebody may switch the supply back on while an electrician is working on the circuit. This could have fatal consequences.

All persons working in the electrotechnical industry are required to be fully competent in carrying out this procedure and students can expect to be asked to demonstrate their ability to do so during practical assignments and assessments. By using this Guide, you will understand how to meet the current requirements of awarding bodies.

2.3.1 What is Guidance Note GS38?

Guidance Note GS38, published by the HSE, contains the recommended requirements for test leads and instruments that are intended to reduce the risk of electrocution, burns or death when carrying out tests on electrical systems.

In some circumstances, this may seem a little over the top – but why risk the possibility that you will potentially suffer a fatal electric shock, when you can use test equipment that has safeguards in place to prevent it?

As well as ensuring that the equipment is safe to use, the person carrying out the test must be adequately skilled to use the equipment.

▼ **Figure 2.3** Test leads

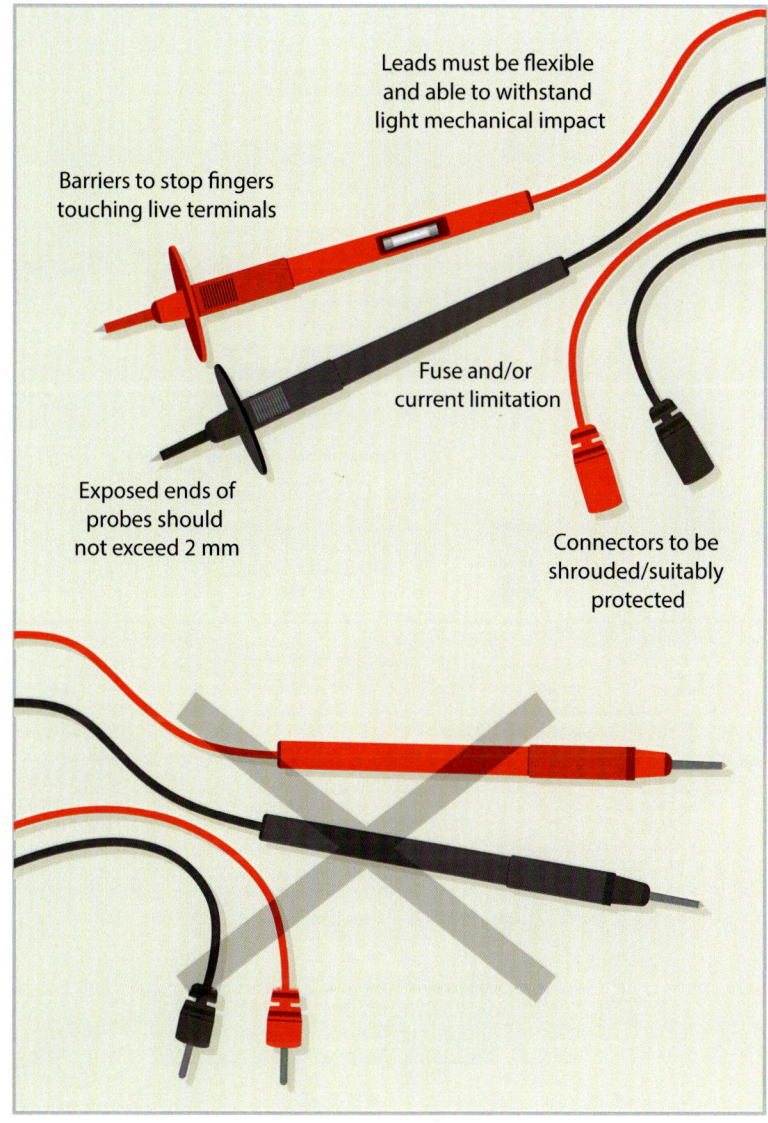

▼ **Figure 2.4** Safe isolation procedure

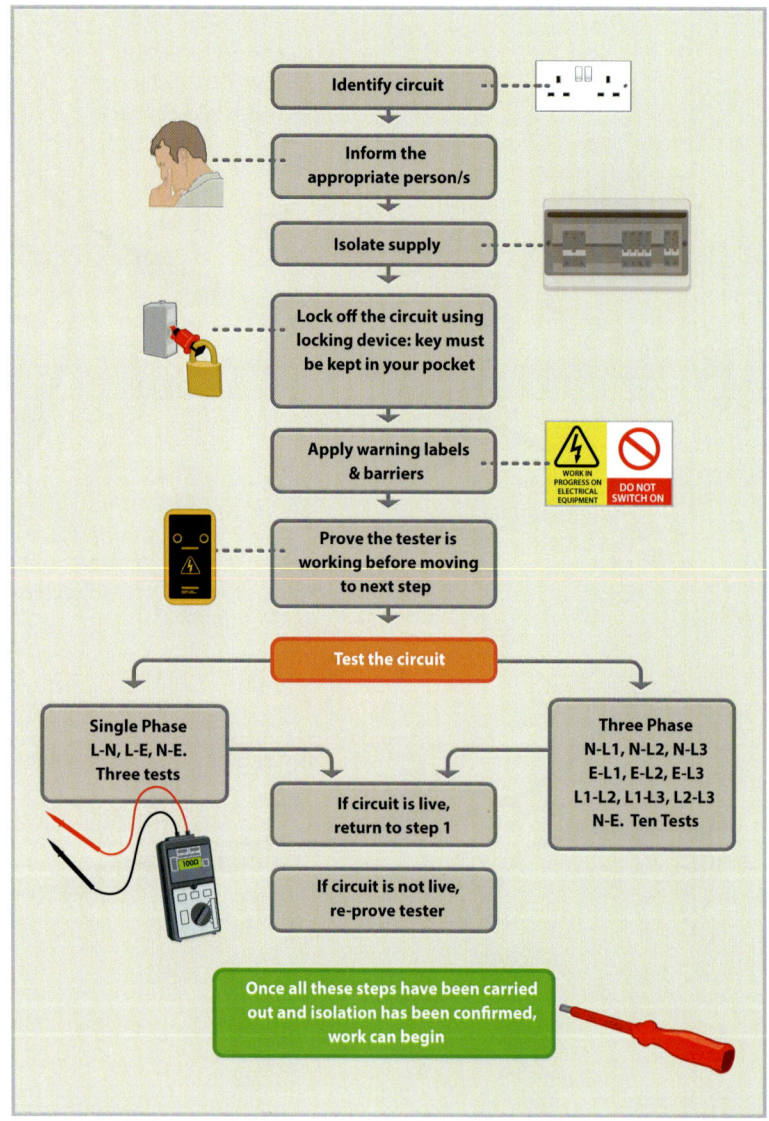

2.4 Tool safety

A basic set of hand tools for an electrician will normally consist of:

(a) a set of insulated (electrical) screwdrivers;
(b) a torque screwdriver;
(c) a pair of side cutters;
(d) a pair of pliers;
(e) a pair of wire strippers;
(f) a hacksaw;
(g) a measuring tape;
(h) a set square;
(i) a ruler;
(j) a file;
(k) a pair of grips; and
(l) a hammer.

All tools have the potential to cause injury if not used correctly. Guidance on how to use the tools is provided by the manufacturer. Training for the use of tools should be provided by instructors and/or employers.

▼ **Table 2.3** Various hand tools used by electricians (Images reproduced with permission from Stanley Tools)

Grips	Grips (also known as 'pipe grips') are fantastic tools, which, when used correctly, are designed to tighten their grip on whatever it is they are turning. It is important to understand that they are designed to turn in one particular direction, which causes them to increase the pressure on what they are holding. If used incorrectly, they will take more effort and will likely slip and potentially injure you.
Hacksaw	A hacksaw or junior hacksaw is a fairly simple, but valuable tool.
	Maintaining a hacksaw is also fairly simple. Make sure it's stored safely and, when the blade gets blunt, replace it. Remember also to lubricate the tightening mechanism. When replacing the blade, ensure that it's fitted the correct way around (if you look closely at the blade, you will see that the points all face one direction; make sure that they are pointing away from the handle when you fit the blade).
	Larger hacksaws usually have adjustable tension. This must be set to a suitable level for the tool to work efficiently.
Hammer	A hammer can be used in a number of situations. Of course, its true purpose is to drive nails into wood or similar material. However, it is known to be used for making minor adjustments to the position of equipment. This is not, however, recommended!
	BS 876:1995 identifies the requirements for hammers.

▼ Table 2.3 *cont.*

Measuring tape 	Not the most dangerous tool in the box, but nevertheless, it needs to be used responsibly. BS 4484-1:1969 identifies the requirements for linear measuring instruments for use on building and civil engineering constructional works. This Standard encompasses steel measuring tapes, steel bands and retractable steel pocket rules. When taking long measurements, the tape can retract quickly and potentially cause cuts if it comes into contact with flesh. Tape measures are susceptible to damage when they are extended beyond the maximum measurement they are designed to take. When this happens, it can be very difficult, if not impossible, to repair them. One key point to remember: measure it twice, cut it once!
Pliers and side cutters 	Side cutters and pliers are probably the most commonly used tools in an electrician's toolbox. As well as cutting a variety of sizes of cable, they are often used to strip off cable sheathing. Pliers can also be used to hold nuts and bolts in place to make it easier to get the thread started. BS 3087-10:1991 sets out the requirements for pliers and side cutters (or 'snips', as they are sometimes called). Just as with screwdrivers, it is important to make sure that the insulation of your side cutters and pliers is in good condition. If there's a breakdown of the insulation on the handle of the tool and it unintentionally comes into contact with a live part, it could present a risk of electric shock to the user.

▼ Table 2.3 *cont.*

Screwdrivers/ torque screwdrivers 	Every electrician will need to have a set of insulated screwdrivers. BS 2559:1994 identifies the requirements for insulated screwdrivers (yes, there is a standard for screwdrivers!).
	This is necessary for safety: if, for example, you were to purchase a screwdriver that did not comply with the Standard, it may not have the required insulating properties and could present a risk of electric shock to the user. Manufacturers of consumer units and similar switchgear assemblies usually require a specific torque setting to be used when terminating conductors.
	To maintain your insulated screwdrivers, it is advisable that you have a suitable means of storing them separately to the rest of your tools. This will avoid them being damaged and the possibility of the insulation failing. If you have a torque screwdriver, it is important to have the settings calibrated in accordance with the manufacturer's instructions.
	If the insulation on your insulated screwdriver is damaged, which can happen when it is used for various other activities (such as removing knock-outs from consumer units or checking for a fixing in ceilings by pushing it through the plasterboard), you must consider replacing the screwdriver, as it may no longer be sufficient to protect you from the risk of electric shock.
Stocks and dies	Stocks and dies are used to thread conduit so that it can be attached by a screwing method to other equipment, such as other pieces of conduit, switch and socket-outlet boxes, etc.
	BS 1813:1974 identifies the requirements for stocks and dies, including the dimensions of conduit dies, die stocks and guides.
Wire strippers 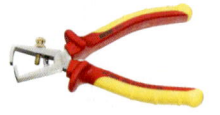	Wire strippers are used to remove the insulation from cable. They can be adjusted to be used on various sizes of cable. Wire strippers can be easily damaged and should be kept separate from heavy tools such as grips, hammers, etc.

▼ **Table 2.4** Various power tools used by electricians (Images reproduced with permission by DeWalt)

110 V drills 	110 V is commonly used on construction sites, although electricians who work in a variety of environments may have a portable 230/110 V transformer so that they can use 110 V in a variety of places.
230 V drills 	Mains voltage drills can be used in most environments by plugging into a 13 A socket-outlet.
Battery drill 	Battery drills are very common in construction. Technology has improved battery life and reduced the weight of such drills significantly. Battery drills can be used for drilling small holes for fixings and access. Some have settings so that they can be used to chase walls (see 'chasing' in Table 1.6). BS EN 62841-2-1:2018+A11:2019 identifies the requirements for the safety of hand-held, battery-powered, motor-operated tools and battery packs, and particularly the requirements for drills.
Chop saw 	These are used to cut conduit, slotted and channel cable tray, trunking and much more. They are fairly heavy-duty cutting tools that go easily through light metal.

▼ Table 2.4 *cont.*

Jigsaw	A jigsaw is used to cut through various types of material, including metal, and is normally used on a flat surface. Electricians might use this type of saw to cut out entry holes in large distribution boards or lengths of trunking.
Threading machine	When installing conduit, the ends must be screwed together and to the equipment. This is normally done by hand with stocks and dies.

2.4.1 Choosing the correct tool for the job

Tools are designed to be used for specific tasks and with specific pieces of equipment. Many manufacturers of tools and components design their products to be used only with their own equipment. This helps to reduce the damage that may be caused by the incorrect use of tools and to maintain a good standard of installation.

Various consequences may result from using the incorrect tools:

(a) Damage to terminal screws

Using the wrong type of screwdriver can cause damage to terminal screws, as they won't fit correctly. It's important to feel if the screwdriver fits the screw snugly and it is therefore necessary to know the differences between slotted, Pozidriv (PZ), Phillips Head (PH), etc. Any excess movement will be liable to cause damage to the screw head as it is tightened.

(b) Undertightening or overtightening terminal screws

If an undersized screwdriver is used, it may be difficult to apply the required torque to ensure a good termination. If an oversized screwdriver is used, it may overtighten the screw and cause damage to the terminal that may cause the conductor to become loose.

(c) Damage to the appearance of accessories and equipment

Equipment is expensive and many electrical items are chosen for their appearance. If the wrong type of tool is used to install the equipment, it is more likely to damage the equipment.

(d) Loose glanding of cables at entry points

When connecting cable glands, it can sometimes be difficult to reach with ordinary tools and specialist equipment must be used. Some glands form part of an earthing system as well as an integral part of a circuit protective conductor (cpc). Not terminating them properly can affect the continuity readings when it comes to testing and can fail to provide a sufficient electrical connection to earth.

(e) Injury to the user

Tools can be dangerous, especially when they aren't used for their intended purpose. For example, using a drill bit instead of a de-burring tool to remove excess material could result in the drill bit breaking, becoming a projectile and causing injury.

(f) Damage to the tools

Tools are expensive. If they are not being used for their intended purpose, they are more susceptible to damage and replacement. Using a large screwdriver as a chisel, for instance, would certainly reduce the life of the screwdriver.

	Test your knowledge
1.	Which guidance published by the HSE contains the recommended requirements for electrical test equipment for use on low voltage electrical equipment?
2.	The insulation on tools designed for electrical work is to protect the user from what?
3.	When safely isolating a circuit, where should the key be kept while work is being carried out?
4.	True or false: Once a risk assessment has been carried out and documented, it does not need to be looked at until work has been completed.
5.	What tool should be used to remove insulation from a cable so that it can be terminated correctly and not damage the conductor?

Generation and Transmission 3

This Section provides information on the following topics:
- ▶ What is required to generate electricity?
- ▶ The characteristics of the electricity generated in the UK
- ▶ How electricity is transmitted
- ▶ How electricity is distributed
- ▶ How electricity is measured

Although the generation, transmission and distribution of electricity are not covered by the IET Wiring Regulations, it is essential to understand these processes in order to appreciate their relevance to electrical installations in general.

3.1 What is used to generate electricity?

Electricity is generated in the UK using various different methods. The most common of these are listed in Table 3.1.

▼ **Table 3.1** Generation sources

Fuel	Type	Environmental impact
Coal	Fossil fuel: non-renewable	High CO_2 emissions
Gas	Fossil fuel: non-renewable	High CO_2 emissions
Oil	Fossil fuel: non-renewable	High CO_2 emissions
Nuclear	Fission: non-renewable	Low CO_2 emissions
Solar	Renewable	Low CO_2 emissions
Wind	Renewable	Low CO_2 emissions
Hydro	Renewable	Low CO_2 emissions
Wave	Renewable	Low CO_2 emissions
Petrol/diesel generator	Fossil fuel: non-renewable	High CO_2 emissions

NOTES:
1 The renewable sources listed in Table 3.1 produce no carbon emissions during their operation; however, the production of the materials involved may have low carbon emissions.
2 The term 'carbon neutral' is often quoted regarding generation of electricity or manufacturing processes and this implies that the CO_2 that is being produced has been balanced out in some way, for example, by the planting of trees that absorb CO_2 and turn it back into oxygen.

3.1.1 Power stations

The basic principles of generating electricity at any power station are mostly the same. Some form of fuel, such as coal or gas, is burned in a furnace to heat water into steam in a boiler. The steam travels through pipes at high pressure and is used to turn a turbine at high speed, which turns a generator. (Steam is used to power a steam train in the same way, but instead of powering wheels on a track, steam in a power station turns a generator.) Generators are made up of coils of copper wire and iron to form electromagnets. When these electromagnets are turned within one another, electricity is created. The electricity generation process is outlined in Figure 3.1.

▼ **Figure 3.1** Generation of electricity

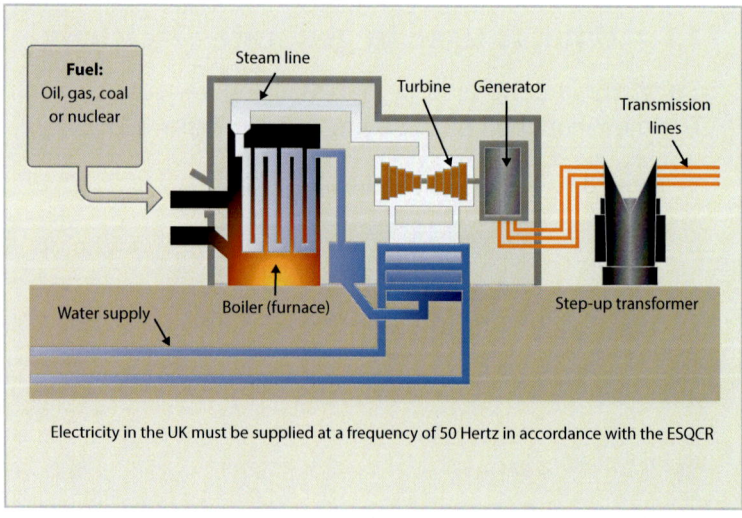

Electricity in the UK must be supplied at a frequency of 50 Hertz in accordance with the ESQCR

3.1.2 Hydro and wind generation

Hydro and wind generation processes use generators that work on the same principle as the ones in power stations, but they are powered by the flow of wind and water instead of pressurized steam.

3.1.3 Solar power

Solar power, on the other hand, works quite differently. A variety of materials are combined, using specialized techniques, to produce what are known as solar panels. These solar panels have the ability to use energy from the sun to create electrical current. By controlling this current, it is possible to harness the energy from the sun and generate electricity.

3.2 What are the characteristics of the electricity generated in the UK?

The ESQCR require the distributors of electricity in the UK to provide users with a standard power supply of similar characteristics across the whole of the UK. There are a number of reasons for this. One good example is electric clocks: in order for a clock to maintain accurate time, the power supply must be at a certain frequency as it basically counts the change of direction of the current used to drive a synchronous motor. For clocks across the country to show the same correct time, the supply frequency must be the same at each clock. In the UK, that frequency is 50 Hz, which means that the alternating current changes back and forth 100 times a second and the low voltage (LV) supply will be 230 V between line and neutral. For three-phase supplies, it will be 400 V between line conductors and 230 V between line and neutral.

The declared nominal supply voltages may vary within permitted tolerances of +10 % and -6 %. 230 V became the UK's nominal voltage for single-phase supplies and 400 V for three-phase in 1995.

3.3 How is electricity transmitted?

3.3.1 The journey of electricity

Electricity is produced by generators at power stations at about 25,000 V. This is not enough to send it long distances along transmission lines, so the electricity needs to be changed through a step-up transformer at the power station, increasing the voltage to between 275,000 V (275 kV) and 400,000 V (400 kV) (see Figure 3.2). Electricity travels long distances more efficiently at higher voltages.

▼ **Figure 3.2** Transmission from power station to the National Grid

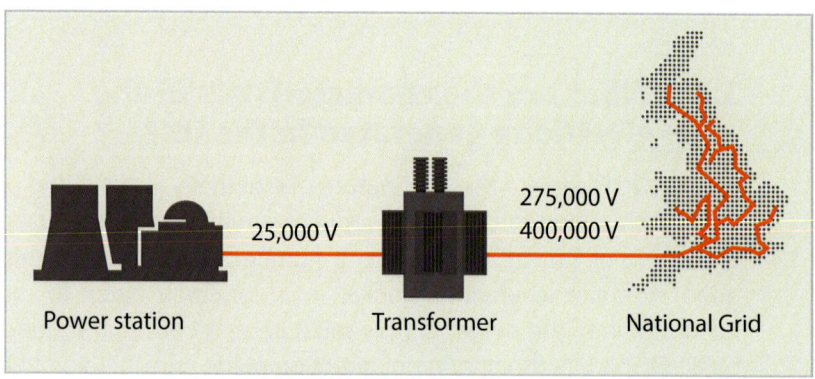

When the electricity leaves the transformer, it travels through the National Grid (sometimes just called the 'Grid'). The Grid is the network of conductors connected to one another across the UK that transmits electricity from power stations to towns and cities. It is designed to be interconnected with every power station in the UK, so that if, for some reason, one power station fails, the towns and cities in that area will still have power. The wires that carry the electricity in the Grid are called 'transmission lines'. These lines are supported across the country by tall electricity transmission towers, often referred to as 'pylons' (see Figure 3.3).

▼ **Figure 3.3** Examples of transmission towers

3.3.2 Transmission

Before electricity from the Grid can be used by factories, shops and homes, it must be transmitted to transmission substations so that the electricity can once again be transformed. This time, however, the process is the other way round: the voltage is much too powerful to use in our homes and businesses and so must be reduced.

Transformers in the substations reduce the very high 275 kV or 400 kV to 132 kV before it enters the regional distribution network. Regional distribution networks carry electricity to distribution substations, where the voltage is again reduced, from 132 kV to 11 kV. The 11 kV network supplies towns, industrial estates and villages, as well as some industrial premises that have large electricity requirements, such as car manufacturing plants.

3.3.3 Distribution

▼ **Figure 3.4** Example of a distribution transformer

The final stage is the local substations and distribution networks. The grey boxes that can be seen in fenced-off areas, with high voltage (HV) signs posted, are the local substations and distribution transformers.

At this stage, the voltage is reduced to 400 V or 230 V. Many commercial outlets, such as shops and supermarkets, will have a 400 V supply, while domestic premises (homes) will normally have a 230 V supply. If you look at the technical information written on your household appliances, such as the kettle and TV, you will see that the voltage will be 230 V.

3.4 How do we measure what we use?

▼ **Figure 3.5** Electricity meters

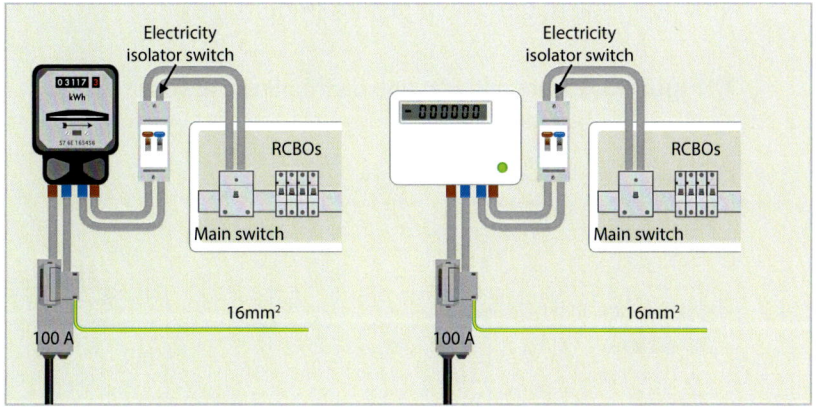

NOTE: RCBO stands for 'residual current circuit-breaker with integral overcurrent protection'. This incorporates both a residual current device, for fault protection, and an overcurrent device in a single unit.

Unfortunately, electricity is not free. In fact, it is reasonably expensive, with the average household spending around £1,000 a year on their electricity bill.

So how is it measured? You will probably be familiar with your consumer unit at home. Next to it, you will generally find some other equipment and thick cable. Your electricity meter will either have a spinning disc behind a glass screen or a little flashing light with an LCD (liquid crystal display) with lots of numbers on it.

The meter records the power used in kilowatt hours (kWh). The electricity supplier will sell the power to the customer for an agreed price. The details of the meter need to be provided to the electricity supplier so that they can accurately charge the customer for what they use.

Smart meters include an 'In Home Display' (IHD), which shows the energy use in terms of cost and consumption and automatically sends meter readings to the supplier every 30 minutes. It provides the consumer with an immediate indication of the energy being used.

3

Whenever electricity is used in a property, it will pass through these meters, recording every little bit of power that is used. Remember that when you leave the bathroom light on next time!

Now let's look at the whole picture (Figure 3.6).

▼ **Figure 3.6** Transmission and distribution network

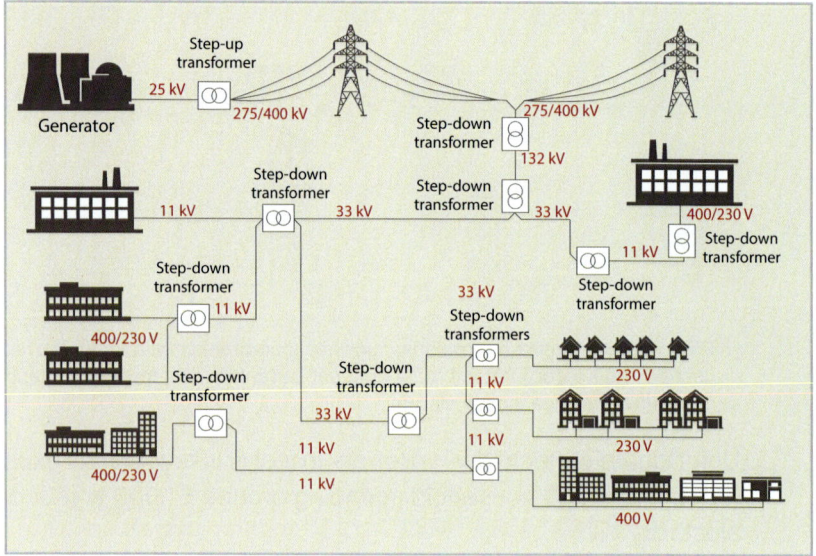

	Test your knowledge
1.	Name three types of supply source which have low carbon emissions.
2.	What does ESQCR stand for?
3.	Why is voltage increased when electricity is being transmitted?
4.	Where might you find a distribution transformer?
5.	What is the frequency that electricity is generated at in the UK?
6.	When electricity is measured at the consumer's installation, what is the unit of measurement used?
7.	What is the greatest benefit of the National Grid?
8.	What does a step-up transformer do?
9.	What does a step-down transformer do?
10.	What are the maximum and minimum single-phase supply voltages within the permitted tolerances in the UK?

Supply 4

This Section provides information on the following topics:
- ▶ Supply intake arrangements
- ▶ Live conductors
- ▶ Current
- ▶ Earthing arrangements
- ▶ Tails
- ▶ Consumer units
- ▶ Ways and modules

4.1 Supply intake arrangements

Before any work is carried out within an electrical installation, details relating to the supply must be looked at and recorded, because such details can change the way in which work is done on the installation and affect what type of protection is necessary. The supply details can be looked at in three ways:

1. live conductors – how many are there?
2. current – what is the maximum rating?
3. earthing arrangement – what type of earthing system is used?

4.2 Live conductors

The supply to smaller domestic and commercial installations is usually 230 V single-phase, meaning that there are two live conductors (line and neutral). Always remember that a neutral carries current when the circuit is in use, so it is classed as a live conductor.

Larger installations, which normally need more than 100 A, will usually be provided as three-phase supply. This allows the larger overall load to be spread over the three-phase supply within the installation, thereby reducing the risk of overloading one phase.

▼ Table 4.1 Live conductor arrangements within the UK

Type	Arrangement	Location
Single-phase 230 V between line and neutral	Two-wire.*	Dwellings, small shops or small commercial installations.
Three-phase 400 V between line and line	Three-wire (no neutral).*	Installations are rarely three-phase only. Often a neutral conductor is required, as single-phase (i.e. line-neutral) circuits will be needed for lighting and appliances.
Three-phase and neutral 230 V between line and neutral; 400 V between line and line	Four-wire.*	Larger retail outlets, larger commercial or industrial buildings.

* An earthing conductor is likely to be present, but it's not deemed to be a live conductor.

4.3 Current

Another factor that affects the supply is the current rating, which is often referred to as 'maximum demand'. Most average-sized houses in the UK will have a maximum demand of less than 100 A and many houses in a road or estate will be supplied from the same substation transformer.

Larger installations that have a maximum demand over 100 A will likely be supplied using three-phase supply, in order to spread the load in the supply. Where a building needs a supply over 400 A per phase, it may have its own substation transformer installed on site.

4.3.1 The distribution network operator (DNO)

A distribution network operator (DNO) is the party licensed to distribute electricity in the UK. They own and operate the system of cables and equipment which brings electricity from the National Grid to installations in homes and businesses.

The maximum demand for an electrical installation will be governed by the rating of the DNO's service cut-out fuse. This means that, where an existing installation is to be added to, you must always check that the supply has enough capacity for the extra load.

Where a new building is to be given a supply, the DNO needs to know the proposed maximum demand for the electrical installation, so that it can provide an adequate supply. It would be the job of the designer to provide this.

Typical current ratings of DNO service fuses are 63 A, 80 A, 100 A, 125 A, 150 A, 200 A, 300 A, 400 A and 500 A.

The DNO's service fuse will be fitted inside the electrical service head (cut-out), which is the enclosure into which the supply cable is terminated before the meter. Most DNO fuses will be in a fuse carrier that is 'sealed' to the service head, meaning that it can only be removed or replaced by a person authorized by the DNO. The seal is a small wire which, with a special sealing crimp tool, locks the fuse to the service cut-out: it has to be broken if the fuse is removed. Connections to meters will also be sealed in this way. If you need to remove a fuse from a service head, you must always contact the DNO.

It is important to know the supply rating, but it is not always easy to see what the DNO fuse rating is. In this situation, you should always contact the DNO, who can tell you all the supply details, including the rating. The DNO has a statutory obligation to provide a designer with details of the supply, including the earthing arrangements.

4.3.2 The distribution system operator (DSO)

The distribution system operator (DSO) is the party responsible for operating the distribution system. Due to the increase in distributed generation on the network and the increase in prosumer's electrical installations, this will require DNOs to take on system operator functions. This includes active network management using latest technology and real-time data to make changes on the network. As a result, a transition from DNO to DSO is taking place and eventually the DNOs will become DSOs.

4.4 Earthing arrangement

The earthing arrangement reflects how the electrical installation and supply is earthed. In the UK, there are three common earthing arrangements, which are described using abbreviations (see Sections 4.4.1 to 4.4.3). The type of earthing arrangement given by the DNO depends a lot on where the installation is located and/or how old the supply cable is.

There are three common arrangements:

1. TN-C-S arrangement;
2. TN-S arrangement; and
3. TT arrangement.

For more information on these, see Section 6.1.4.

4.4.1 TN-C-S arrangement

The DNO earth and neutral are combined in the same conductor of the supply cables. In this arrangement, as the supply neutral is also the earth, it is called a PEN (protective earthed-neutral) conductor. Once the electrical supply is terminated into the service head, the PEN conductor is split into two terminals: one is neutral, the other is earth.

This is the most common earthing arrangement in the UK. It is also called PME (protective multiple earth), because it could have more than one 'source' earth electrode along the length of the buried supply cable in the street. For this system, the DNO will declare an expected maximum external earth fault loop impedance (Z_e) of 0.35 Ω.

This system may be identified by looking at the DNO service head. If the earthing conductor comes from inside the service head, it is likely to be PME. (If in doubt, the DNO should be contacted for clarification.)

▼ **Figure 4.1a** TN-C-S arrangement

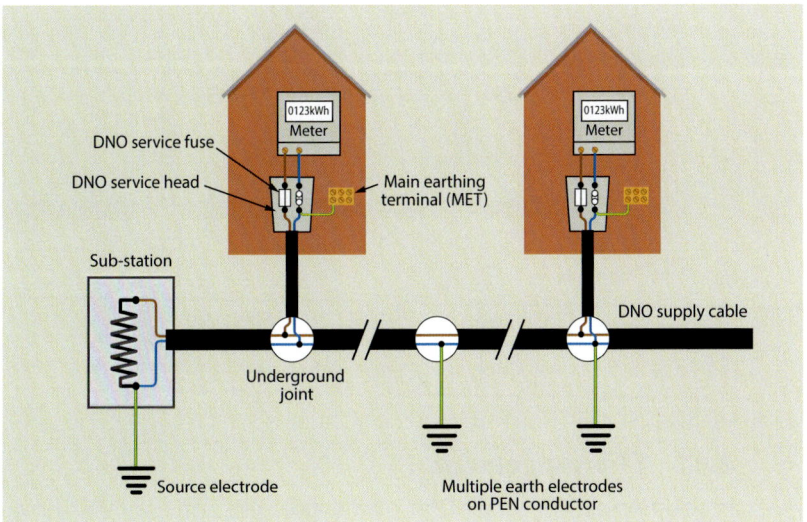

4.4.2 TN-S arrangement

The DNO earth and neutral are separate conductors in the supply and installation. The supply earth is normally the metallic sheath or armour of the supply cable.

This arrangement is common in older houses. For this system, the DNO will, in most circumstances, declare an expected maximum external earth fault loop impedance (Z_e) of 0.8 Ω.

This system may be identified by looking at the supply cable below the service head – the earthing conductor is often clamped to the metallic sheath or armour of the supply cable. (If in doubt, the DNO should be contacted for clarification.)

▼ **Figure 4.1b** TN-S arrangement

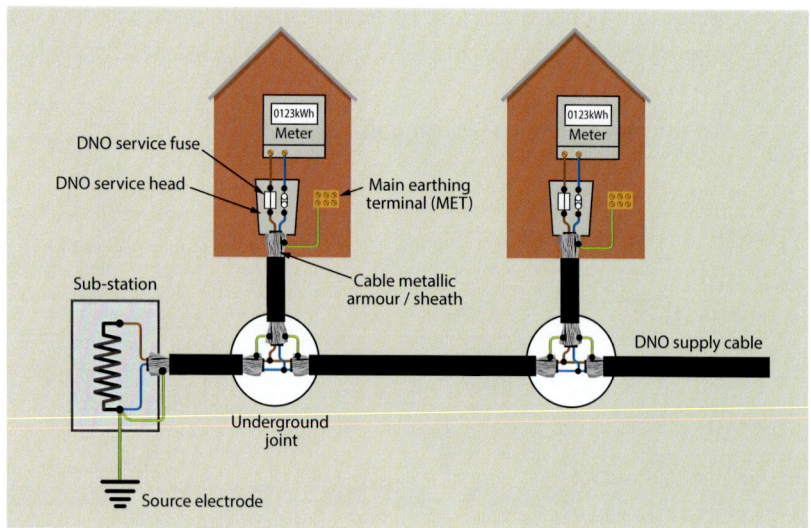

4.4.3 TT arrangement

In some cases, the DNO will not provide an earth and the responsibility lies with the consumer to provide this in the form of an earth electrode. The two-wire supply to this type of installation will normally be carried via overhead cables.

This arrangement is more common in rural areas, where transformers are mounted on poles serving, for example, one or two dwellings. More and more of these are being converted by the DNO to TN-C-S arrangements.

This system is identified by the earthing conductor being connected to an earth electrode in the ground. External earth fault loop impedance (Z_e) can vary greatly, depending on many factors, but values of less than 200 Ω are regarded as stable values of Z_e. As the ground is the conductor between earth electrodes, problems can arise when the ground freezes in winter or dries out in summer, so keeping Z_e at less than 200 Ω improves stability.

▼ **Figure 4.1c** TT arrangement

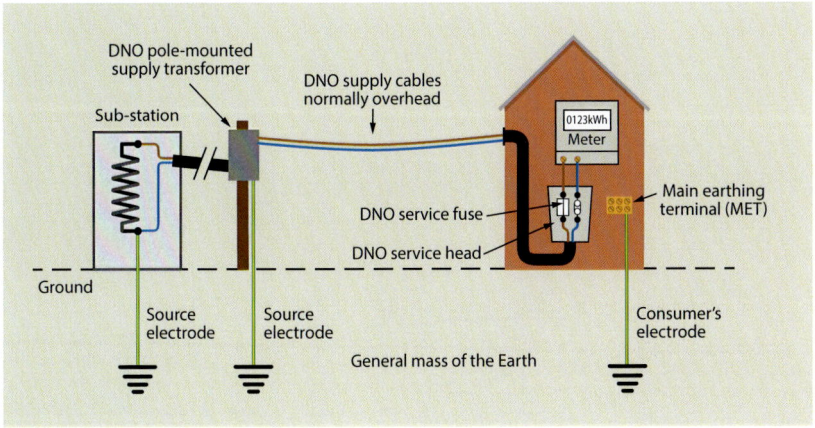

4.5 Responsibility

Once the electrical supply has been metered, it generally stops being the responsibility of the DNO or the supplier (the party selling the electrical energy) and becomes the responsibility of the consumer. The consumer is the person who owns the electrical system on the property – in simple terms, the person who owns the electrical installation. It is important to understand that the supplier and the DNO are not the same.

Depending on where the meter is located, what happens next can vary. In some situations – more likely, in newer buildings – there may be an isolation switch between the meter and the consumer unit. This isolation switch may have been fitted by the DNO or by the electrician who carried out the installation work.

If you are fitting a new installation, it is always a good idea to fit an isolation switch between the meter and the consumer unit, so that work can be carried out on all parts of an installation, including inside the consumer unit with the supply completely isolated.

Isolation switches combined with a fuse (switch fuses) are also fitted where the consumer unit is located some distance away from the meter location.

▼ **Figure 4.2** Typical intake layouts and responsibilities

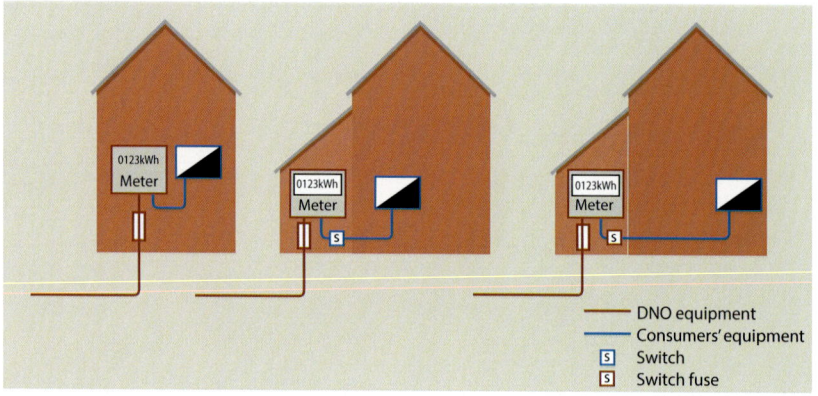

4.6 Tails

Tails is the name given to the single-core, insulated and sheathed cables that go into and out of the meter. Two terms are used:

1. meter tails: connecting the distributor's cut-out to the meter. These are usually owned by the meter operating company.
2. consumer's tails: connecting the meter to the consumer unit or distribution board. These are the responsibility of the installation owner.

Single-core cable is used to give flexibility and allow easy installation. The cable has insulation (the brown or blue covering of the cable), which provides basic protection against electric shock. The cable must also have an outer sheath, which provides mechanical protection against physical damage (this is usually the outer grey covering of the cable).

Tails should be as short as possible, to avoid damage, and should be clipped or installed in trunking to prevent them from being moved. When a cable is moved, this can put stress on the termination and, over time, cause the termination to deteriorate (i.e., the cable can become loose in the terminal). The sheathing must cover the entire cable, especially when inside a metal consumer unit, right up to the point of connection to the main switch inside the consumer unit.

If the consumer unit is located away from the meter point, a much more robust cable, such as steel-wire armoured (SWA) cable, should be used instead of insulated and sheathed tails and overcurrent protection will be required for the cable.

4.7 The consumer unit

Generally, the first unit in an installation, where the supply is split into circuits, is called the consumer unit. Regulation 421.1.201 requires all consumer units in domestic dwellings to be fabricated from a non-combustible material. The consumer unit will have protective devices inside. These could be:

(a) circuit-breakers;
(b) circuit-breakers and residual current circuit-breakers (RCCBs);
(c) residual current circuit-breakers with integral overcurrent protection (RCBOs); or
(d) fuses.

These protective devices protect the circuits. The consumer unit will also have a main switch that can be used to isolate the entire installation. In a domestic dwelling, the main switch must be double-pole, meaning that it isolates both line and neutral. In some three-phase installations that are part of a TN earthing arrangement, the main switch may be three-pole, meaning that it switches the three line conductors only. In this situation, the neutral can be isolated by a link in the consumer unit.

Three phase installations having a TT earthing arrangement must have four-pole switches isolating all live conductors.

In a domestic dwelling, the main switch in a consumer unit must be the type that is rated to switch the full load current. Some isolator devices that act as main switches are not rated for full load current and, if they are used regularly to switch full load, they will eventually fail.

You may hear a consumer unit called many different things, such as 'fuse board', 'distribution board' (DB) or even 'fusebox'. You may also see abbreviations such as CCU and C/U used to refer to consumer units. A DB is often the name given to further units on the end of a distribution circuit. A distribution circuit is a circuit that supplies a distribution board, whereas a final circuit is one that supplies current-using equipment such as lights and heaters or socket-outlets.

▼ **Figure 4.3** Distribution and final circuits

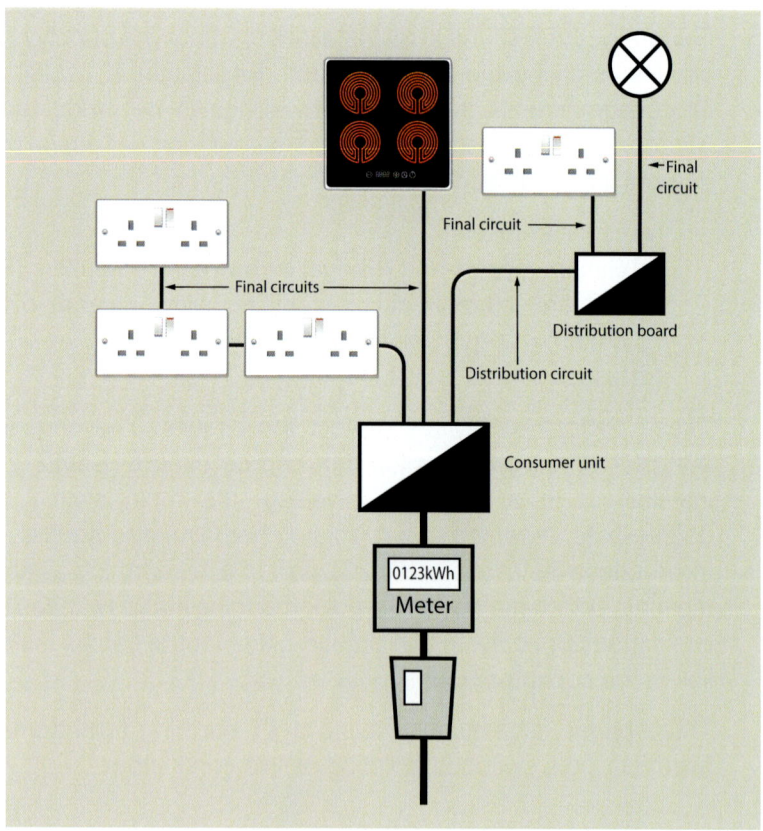

4.7.1 Ways and modules

When you buy a consumer unit, the amount of circuit-breakers or main switches that can be installed are described differently.

(a) **Modules:** this is the number of components that can be fitted into the consumer unit. For example, a main switch controls both line and neutral, so is classed as a two-module device. A circuit-breaker is usually single-pole, so would be one module. If you needed a consumer unit to control six circuits and have a main switch, you would require an eight-module board.

(b) **Ways:** this describes the number of circuits that can be controlled. A six-way single-phase board could house six circuit-breakers as well as the main switch. A six-way three-phase board can house six three-phase circuit-breakers or 18 single-phase circuit-breakers. It could also house a combination of the two, such as two three-phase (six ways) and 12 single-phase (12 ways).

▼ **Figure 4.4** Six-way consumer unit with eight modules

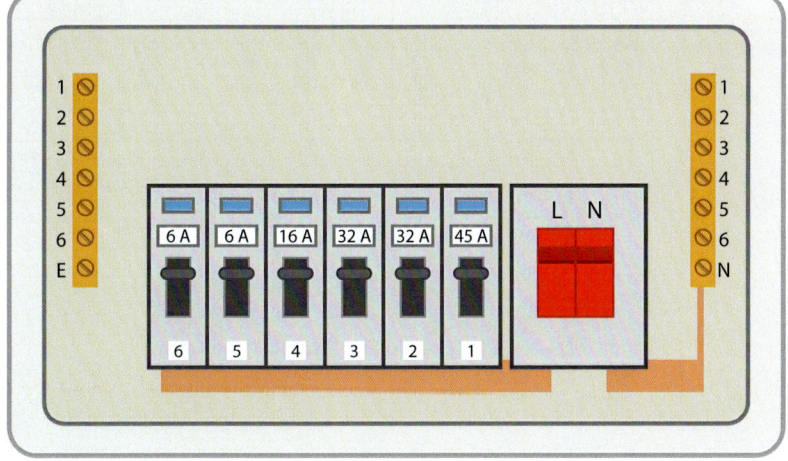

4.7.2 Split-way

Some consumer units are known as split-way boards. This means that some of the modules can be protected by one residual current device (RCD) and the others by a second RCD. Split-way boards should always have one main switch that can isolate all circuits.

▼ **Figure 4.5** Split-way board with three ways protected by one RCD and three ways protected by another (12 modules in total)

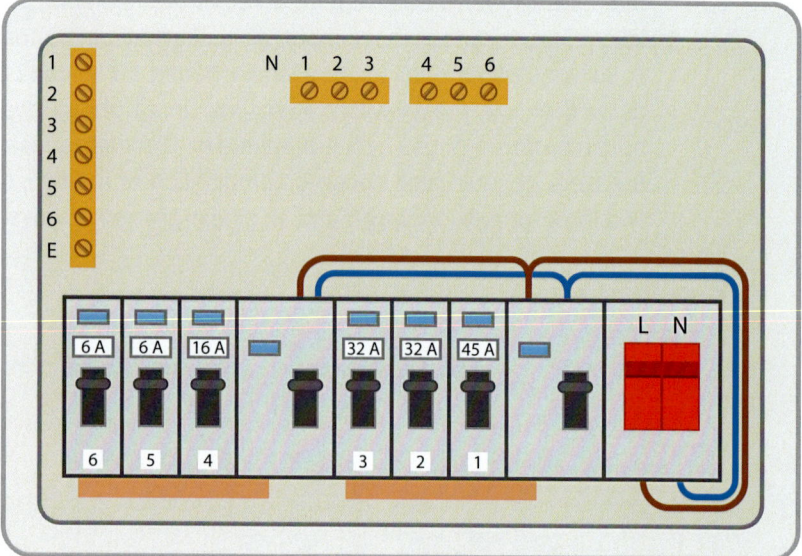

4

	Test your knowledge
1.	What two types of connection are described for a three-phase, three-wire connection according to Part 3 of the IET Wiring Regulations?
2.	What does the term PEN stand for with regard to TN-C-S earthing arrangements?
3.	List the information that a supplier must provide on request in relation to the characteristics of a supply in accordance with Section 313 of the IET Wiring Regulations.
4.	What type of cable should be used for the 'tails' in a domestic property where the cables are installed in free air with no containment system?
5.	Which regulation in Section 421 of the IET Wiring Regulations states that a consumer unit within a domestic (household) installation should have an enclosure made from non-combustible material?
6.	List three reasons given in Part 3 of the IET Wiring Regulations for dividing an installation into circuits.
7.	Six circuits are to be supplied from a consumer unit that incorporates a double-pole main switch. What is the minimum number of modules required in the consumer unit?
8.	Which regulation from Section 536 of the IET Wiring Regulations states that devices installed within an assembly must have their compatibility verified?
9.	What is the maximum rating of a socket-outlet that can be used as a functional switching device according to Table 537.4 of the IET Wiring Regulations?

Protection and Isolation 5

This Section provides information on the following topics:
- ▶ Types of protective devices
- ▶ Uses of protective devices
- ▶ Selectivity
- ▶ Isolation and switching
- ▶ Devices used for isolation and switching
- ▶ Reasons for isolation and switching

5.1 Protective devices

A protective device can provide protection against the various types of overcurrent, such as overload, short circuit and earth fault. An **overcurrent** is any current that exceeds the current-carrying capacity of the circuit conductors. Some protective devices can provide protection against very small earth leakage currents (often referred to as **protective conductor current**). To understand how these devices operate, we first need to define the types of overcurrent.

(a) **Overload:** a current that is more than the normal load current for fairly short durations. There is no fault in the circuit, just too much equipment being used at once.

(b) **Short circuit:** where a fault happens between live conductors, such as line-to-neutral or line-to-line. The fault is assumed to be of negligible or very low impedance, which would develop high fault currents.

(c) **Earth fault:** where a fault happens between line and earth, creating an electric shock risk. The fault is assumed to be of negligible or very low impedance, which would develop high fault currents.

(d) Neutral fault: where a fault occurs between neutral and earth. These often go undetected, but can cause residual current devices (RCDs) to operate.

In addition to the above, protective devices also have different ratings. These are:

(a) Rated short-circuit capacity (breaking capacity) (I_{cn}): this is the amount of current that a fuse can take before it explodes or seriously damages the carrier and its surroundings. In the case of a circuit-breaker or residual current circuit-breaker with integral overcurrent protection (RCBO), faults higher than their rated capacity could weld the contacts together, meaning that the current will not be stopped from flowing.

(b) Nominal rating (I_n): this is the level of current that the device can handle in continued service. For example, a 32 A circuit-breaker can easily carry 32 A load current for the lifetime as specified by the manufacturer without deterioration or loss of function.

(c) Current causing effective disconnection (I_a): this is the value of current that would cause the device to disconnect in the time required. For example, a BS 88-2 fuse with a 6 A rating will disconnect in 0.4 seconds (s) at 18 A. We can determine this using the time/current graphs in Appendix 3 of the IET Wiring Regulations or from the manufacturer's data.

▼ **Figure 5.1** Excerpt from the time/current graph as shown in Appendix 3 of BS 7671

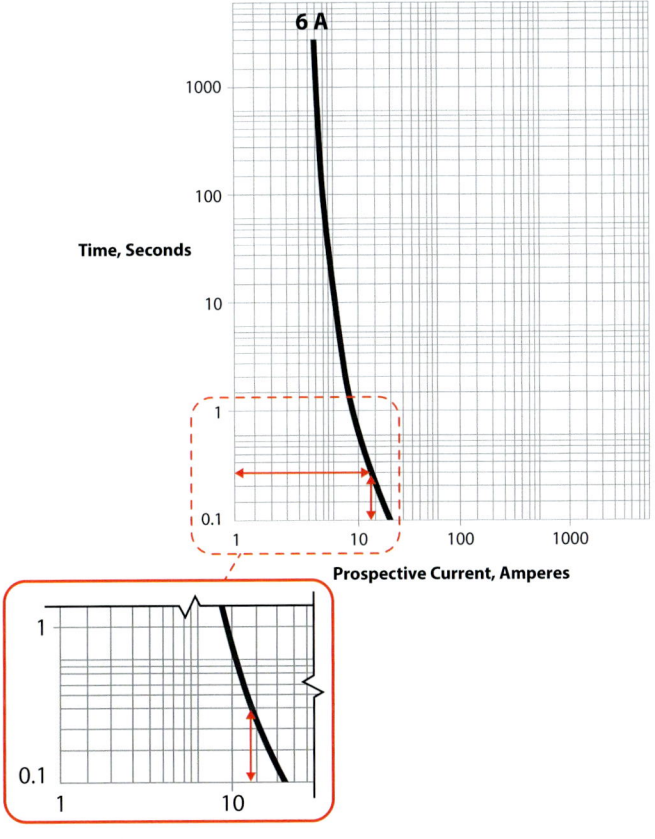

Using the graph, you can see that the 6 A device requires approximately 18 A to disconnect in 0.4 s.

You must always remember that a protective device such as a circuit-breaker may have a nominal rating of 16 A, but it will never disconnect at 17 A or 18 A. In the case of a Type B 16 A circuit-breaker, an approximate current value of 23 A will probably operate the device and disconnect the circuit in around three hours, but 80 A or more is required to disconnect it quickly.

The following table sets out the common protective devices used in many electrical installations, as well as the relevant Standards.

▼ **Table 5.1** Protective devices and relevant Standards

Protective device	Standard
High rupturing capacity (HRC) fuses (also called high breaking capacity (HBC) fuses)	BS 88-2
Cartridge fuses	BS 88-3
Semi-enclosed, rewireable fuses	BS 3036
Plug fuses	BS 1362
Circuit-breakers	BS EN 60898
Residual-current circuit-breaker (with overcurrent protection) (RCBO)	BS EN 61009
Residual current device (RCD)	BS EN 61008

Each type of device is different in the way it performs and handles faults or overloads and a designer will select a particular device based on those differences.

5.1.1 High rupturing capacity (HRC) fuses

As the name of these fuses (also known as high breaking capacity (HBC) fuses) suggests, they can handle very high fault currents, as they have a high breaking (rupturing) capacity. Inside the porcelain body of the fuse, the fuse element is surrounded by sand. When a large fault current causes the fuse element to blow, the heat from the blast turns the sand into glass as it absorbs most of the energy. Because these fuses can absorb a great deal of energy, they can safely handle fault currents up to 80 kA (80,000 A).

These fuses may be classed as 'System E', meaning that they are bolted in place, or 'System G', meaning that they do not have the end fixing lugs and are clipped into place.

These fuses come in a wide range of nominal ratings (I_n), from 3 A up to 500 A and even higher.

There are two protection categories of HRC fuse:

1 gG fuses: these are general use fuses for standard circuit protection. They will operate at lower currents, typically around seven times their nominal rating.
2 gM fuses: these are motor-rated, allowing for a surge current common to motors or discharge lighting starting. These will need higher fault currents to operate.

HRC fuses are normally found in large commercial or industrial buildings.

▼ **Figure 5.2** HRC system E fuse

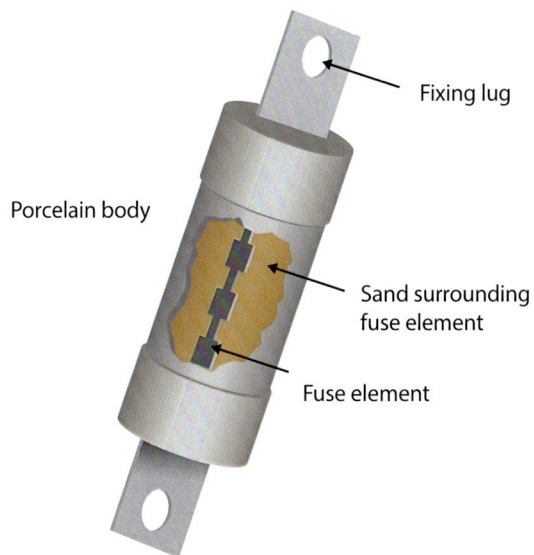

5.1.2 Cartridge fuses

These fuses are similar to HRC fuses, but are intended for use by ordinary persons, so are more common in domestic dwellings. You may come across them in older consumer units in houses and they may be marked as BS 1361, which was the old British Standard number for them.

They generally have a breaking capacity of 16 kA and the nominal ratings (I_n) range from 5 A to 100 A.

Before circuit-breakers were commonly found in dwellings, these fuses were seen as better alternatives to rewireable fuses. In older electrical installations you may therefore see these fuses alongside rewireable fuses in the same consumer unit.

▼ **Figure 5.3** Cartridge fuse

5.1.3 Plug-top fuses

You will find these BS 1362 fuses in every 13 A plug-top or fused connection unit. They range from 1 A to 13 A, but the most common ratings are 3 A, 5 A and 13 A. Their main purpose is to protect an appliance's flexible cable.

▼ **Figure 5.4** Plug-top fuse

5.1.4 Semi-enclosed fuses

Commonly called rewireable fuses, these devices are much less common today, but they are still recognized by the IET Wiring Regulations as suitable. They used to be the most common type of fuse in houses and commercial installations, but are being phased out of use in preference to circuit-breakers, because:

(a) semi-enclosed fuses have a low breaking capacity (typically 1-1.5 kA);
(b) semi-enclosed fuses may contain asbestos in the fuse carrier;
(c) semi-enclosed fuses are not very user-friendly when the wire needs replacing; and
(d) it is very easy to put the wrong fuse wire rating in – for example, a 30 A wire into a 5 A fuse carrier for a 5 A circuit.

▼ **Figure 5.5** BS 3036 semi-enclosed fuse

5.1.5 Circuit-breakers

Circuit-breakers are probably the most common protective devices. They are user-friendly and reliable. 'Circuit-breaker' is the term covered by BS EN 60898; however, manufacturers will still refer to them in three general terms:

1 **circuit-breakers (CB):** these devices to BS EN 60898 come with nominal ratings between 2 A and 125 A. You will see them in all new installations.

2 **miniature circuit-breakers (MCBs):** these devices to BS 3871 have been superseded by BS EN 60898 devices, but you will still see them on many older installations. They come with nominal ratings between 3 A and 80 A and are commonly used in many types of installations as protection for final circuits. The word 'miniature' was dropped when BS EN 60898 superseded BS 3871.

3 **moulded case circuit-breakers (MCCBs):** these devices to BS EN 60947 are larger circuit-breakers, generally used in the large panel boards that protect distribution circuits.

Instead of a breaking capacity, circuit-breakers have two ratings:

1 **I_{CS}:** this is the value of fault current that can be safely handled by a circuit-breaker without it being damaged, and with it remaining usable (serviceable) after the fault.

2 **I_{CN}:** this is the value of fault current that can be safely handled by a circuit-breaker, after which it will not remain serviceable and will need replacing. Any fault current above the I_{CN} rating will be dangerous and could cause either the circuit-breaker to explode or the contacts to weld together, meaning that the fault current will not be disconnected.

For example, a circuit-breaker may have an I_{CS} of 6 kA and an I_{CN} of 10 kA. From Figure 5.6 below, we can see that a fault current up to 6 kA will be safely interrupted by the protective device, which remains usable afterwards. Any fault between 6 kA and 10 kA will be safely interrupted by the protective device, but the device will need to be replaced after the fault, as it may be damaged inside. Any faults above 10 kA could either cause the protective device to explode or weld the contacts together.

Figure 5.6 Relationship between I_{CS} and I_{CN} for circuit-breakers

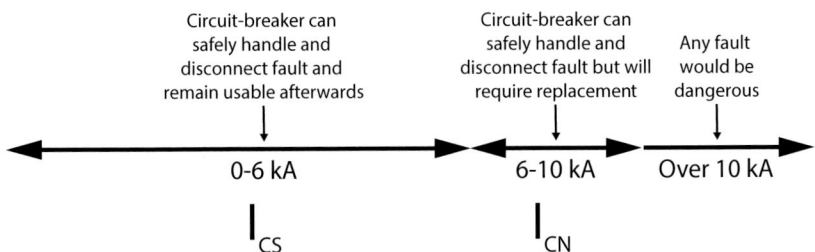

Operation of circuit-breakers

Circuit-breakers operate in two ways:

1. **via a magnetic trip:** this is a solenoid within the circuit-breaker where a pre-set value of overcurrent will create a magnetic field strong enough to disconnect the device instantly (within 0.1 s). The magnetic trip is identified as component 1 in Figure 5.7.
2. **via a thermal trip:** this is where an overcurrent heats up a thermal device, such as a bi-metallic strip, causing disconnection. This process takes longer, depending on the value of the overcurrent. The thermal trip is identified as component 2 in Figure 5.7.

▼ **Figure 5.7** Internal components within a circuit-breaker

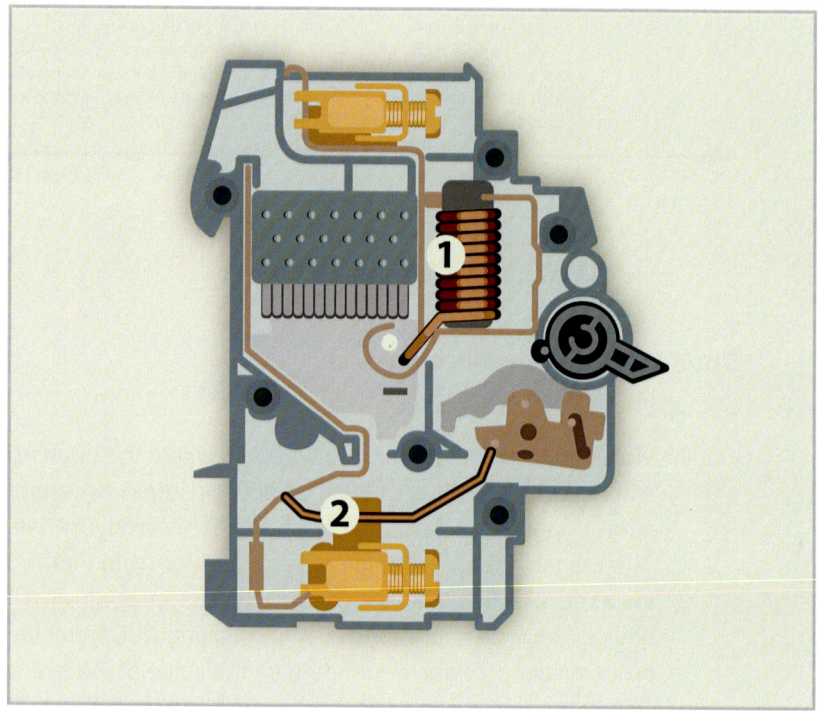

These two ways of tripping a circuit-breaker can be seen in the time current/ graphs for circuit-breakers in Appendix 3 of the IET Wiring Regulations.

▼ Figure 5.8 Excerpt from the time/current graphs for circuit-breakers in Appendix 3 of the IET Wiring Regulations

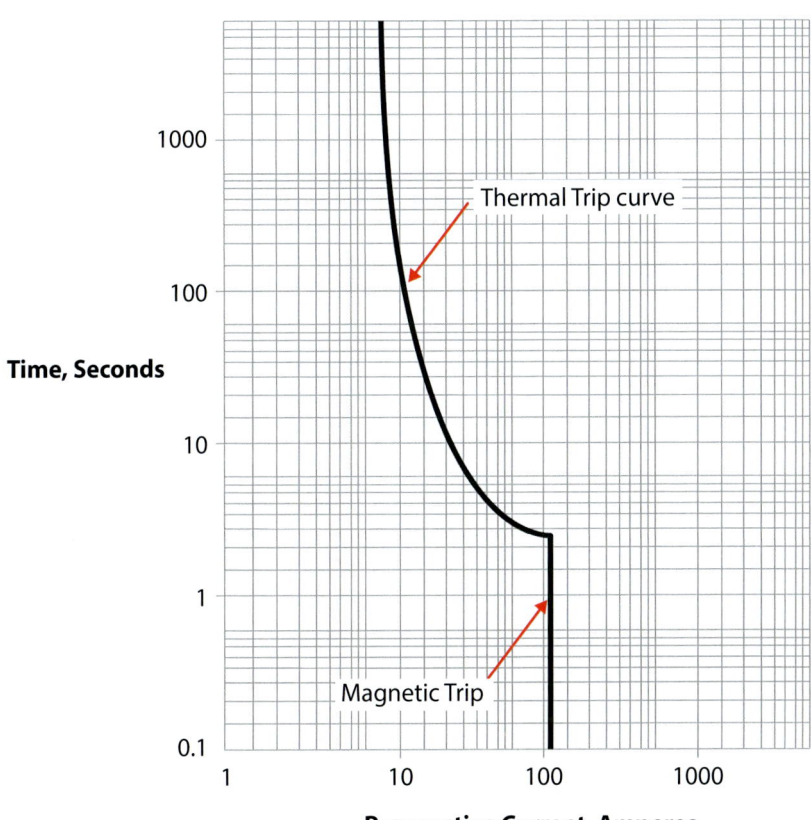

Types of circuit-breaker

There are three types of circuit-breaker. The value of the magnetic trip will vary, depending on the type selected. Some loads, such as fluorescent luminaires, draw a surge current when they first start up. This surge current could be seen as a fault by a Type B circuit-breaker, so a Type C would commonly be used instead, as the magnetic trip for this type is set higher, meaning that more current is required to instantaneously trip the device. Because the thermal trip takes longer to react, the circuit-breaker doesn't trip with the surge current. This surge current should also be taken into consideration when selecting the switches for the circuit, so that they can be selected with the appropriate rating.

The three types of circuit-breaker are:

1. **Type B:** the magnetic trip is set to approximately five times the nominal rating, so a device that has a nominal rating (I_n) of 10 A will trip instantly (within 0.1 s) with a fault current exceeding 50 A. This type would typically be used to protect general circuits, such as socket-outlets or fixed heating appliances.
2. **Type C:** the magnetic trip is set to approximately 10 times the nominal rating, so a device that has a nominal rating (I_n) of 10 A will trip instantly with a fault current exceeding 100 A. These devices would be used to protect loads that surge on starting, such as discharge lighting or motors.
3. **Type D:** the magnetic trip is set to approximately 20 times the nominal rating, so a device that has a nominal rating (I_n) of 10 A will trip instantly with a fault current exceeding 200 A. This type should only be used on very specialist loads, such as welding sets that create high current surges for short durations.

5.1.6 Residual current devices (RCDs)

RCDs are different from circuit-breakers as they do **not** provide short-circuit or overload protection. Instead, RCDs monitor the current passing through the device into the circuit (line) and the value coming back (neutral). If these values are different, some current must be escaping to earth. If the difference between the values on either side, i.e. line and neutral, is bigger than the residual operating current setting (I_n), the device will trip.

▼ Figure 5.9 Typical RCD arrangement

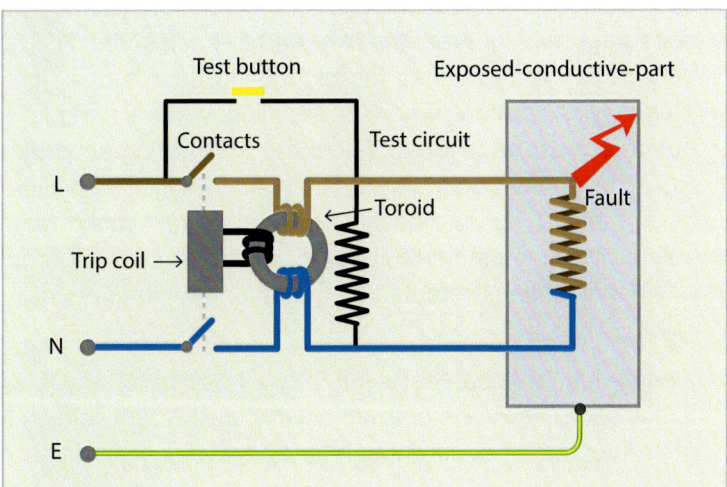

RCDs should be rated in two ways:

1. **nominal current:** this is based on the load current in the circuit, as it is the amount of current that the device can safely switch. In the case of a circuit with a load current of 17 A, an RCD with a nominal rating of 20 A would be used.
2. **residual operating current setting ($I_{\Delta n}$):** this is the value of imbalance current, in milliamps (mA), that causes the device to trip within 300 milliseconds (ms). RCDs are available in a range of residual operating current settings, such as 10 mA, 100 mA, 300 mA, 500 mA and, most commonly, 30 mA.

Where RCDs are used in series, as in a TT installation, for example, it is important that in the event of a fault to earth, the RCD closest to the fault, on the supply side, operates first. This ensures that there is minimum disruption of the supply for any fault to earth which occurs. In the TT installation, an 'S-type' RCD would be used at the origin, with a time delay, and a general RCD used to protect each final circuit. The 'S-type' time delay would be set to allow the general RCD to operate first in the event of a fault to earth on a final circuit and thereby maintain the supply to the rest of the installation. This arrangement is referred to as 'selectivity'.

RCDs are found in many forms: they can be built into accessories such as socket-outlets and fused connection units, incorporated into consumer units or mounted in separate enclosures. RCDs can even be part of main switches for distribution boards or consumer units. This may be common where an installation forms part of a TT system, but care must be taken to select the RCD rating accordingly, as the RCD must be capable of use as a main switch. The device must also be of the 'S-type' and have a residual current rating that allows for selectivity, as a single fault could otherwise cause the power in the entire installation protected by that main switch to be lost.

RCDs, unlike circuit-breakers, have a button on them labelled 'test'. If something mechanical is left in one position for any length of time, it may stick. As RCDs operate at very low current values, the energy generated may not be enough to trip a sticky mechanism.

Because of this, an RCD test button must be pushed regularly to keep the mechanism free from sticking. Circuit-breakers do not need this button, as they operate at high current values, which will trip a sticky mechanism.

5.1.7 Residual current circuit-breakers with integral overcurrent protection (RCBOs)

Put simply, an RCBO is a circuit-breaker that also has an RCD built into it, giving it overload, short-circuit and residual earth current protection. RCBOs fitted for circuit protection require a neutral connection as well as the line, because they need to monitor the individual circuit current for imbalances.

RCBOs are able to meet the requirements of the IET Wiring Regulations in terms of additional protection and protection against overcurrent, as well as meeting the requirements for independent circuit control (where a fault on one circuit does not affect another circuit by disconnecting it).

Like RCDs, RCBOs have a test button that requires regular use, for example, by being pressed every six months.

5.2 Selecting protective devices

▼ **Table 5.2** Protective devices

Device type	Advantages	Limitations	Common use
BS 88-3 fuse	A reasonable alternative to a rewireable fuse without having to change the consumer unit. High breaking capacity.	Fairly poor overload protection. Separate RCD needed to meet additional protection requirements. Requires replacement following rupture.	Used to be referred to as BS 1361. Found in older dwellings.
BS 88-2 gG & gM fuses	Very high breaking capacity. Able to select gM rating for motors or discharge lights.	Requires replacement following rupture. Separate RCD needed to meet additional protection requirements.	At the origin of installations that have a high prospective fault current. In industrial switch fuses.
BS 3036 fuse	Inexpensive, and has the ability to determine the nature of the fault, as you can visibly see the broken fuse wire. Low breaking capacity.	Requires replacement following rupture. Separate RCD needed to meet additional protection requirements. Wrong rating of wire could easily be used.	Found in older dwellings and some old commercial/industrial applications. Not generally used for new installations. BS 7671 still permits their use, but certain conditions must apply, such as de-rating circuits and low prospective fault currents.

▼ Table 5.2 *cont.*

Device type	Advantages	Limitations	Common use
BS EN 60898 circuit-breaker	Good overall circuit protection and easily reset following fault or overload. Good range of types, depending on load.	Easily reset, meaning that ordinary persons could continually reset them, causing faults to worsen and shortening the life of the circuit-breaker. Have low short-circuit capacity compared with other devices.	Found in all new installations: dwellings, commercial and industrial.
BS EN 61009 RCBO	Ability to meet all the requirements of circuit protection.	Breaking capacities not suitable for very high prospective fault currents.	Where circuits require independent circuit protection meeting all requirements of BS 7671 (depending on prospective fault current values).
BS EN 61008 RCCB	Provides residual earth fault current protection where needed by special locations or additional protection. Can be installed remotely from an older consumer unit so an existing consumer unit can remain in place.	Does not provide short-circuit or overload protection.	Where additions or alterations are made to installations protected by fuses or as main or split-load protection.

5.3 Selectivity

In the IET Wiring Regulations, the term 'selectivity' is used and is defined as the ability of a protective device to operate in preference to another in series with it.

Good selectivity is important when designing an electrical installation that contains many protective devices. Selectivity essentially means that the protective device nearest to the fault will operate first. If a fault in an appliance's flexible cable happens, ideally, the fuse in the plug should blow before the circuit-breaker/RCBO that protects the circuit. This is good selectivity. If the circuit-breaker/RCBO operates first, power will be lost to all appliances on that circuit. This is an example of poor selectivity.

Equally, if a fault occurs in a circuit and the DNO's service fuse operates, poor selectivity will mean that the entire building will lose power.

Just because certain devices may have nominal current ratings (I_n) higher than others in the same circuit, it doesn't always mean that good discrimination has been achieved. The type of circuit-breaker also plays a part, as well as the type of fault.

Designers can check selectivity using the time/current graphs found in Appendix 3 of the IET Wiring Regulations or provided by manufacturers.

Looking at the two examples below, the first graph shows that the characteristic lines for the local circuit-breaker and the distribution circuit fuse do not cross at any point. This means that any fault in the circuit will trip the circuit-breaker first.

The second graph shows the local circuit-breaker line crossing the distribution fuse. This means that fault currents higher than the point where the lines cross will cause the distribution fuse to disconnect first.

▼ **Figure 5.10** Examples of selectivity

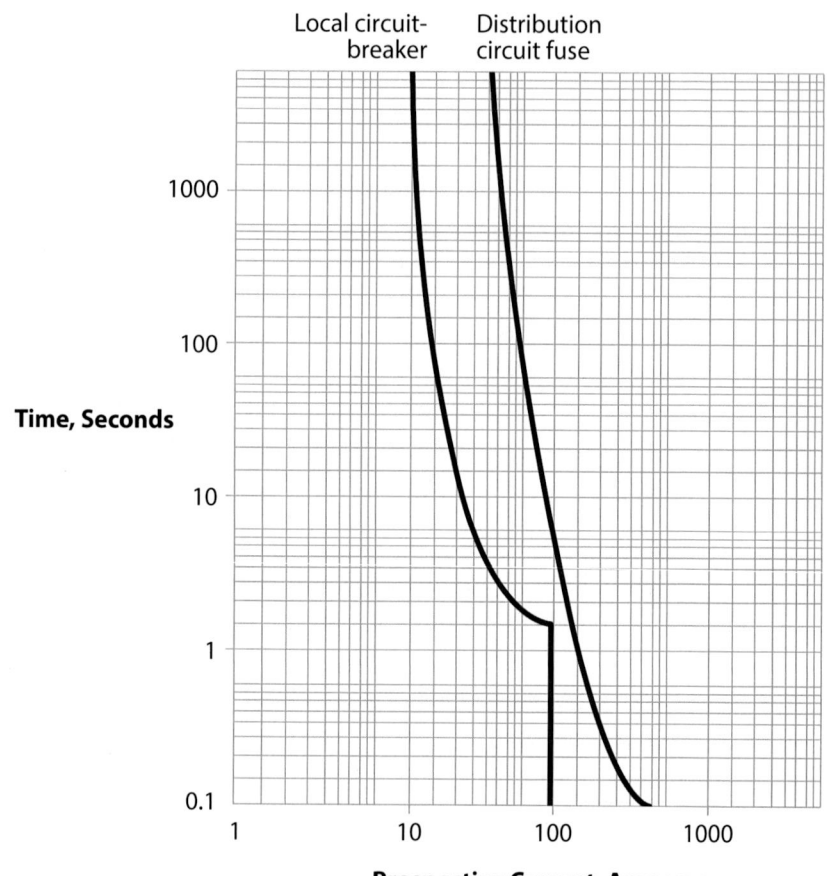

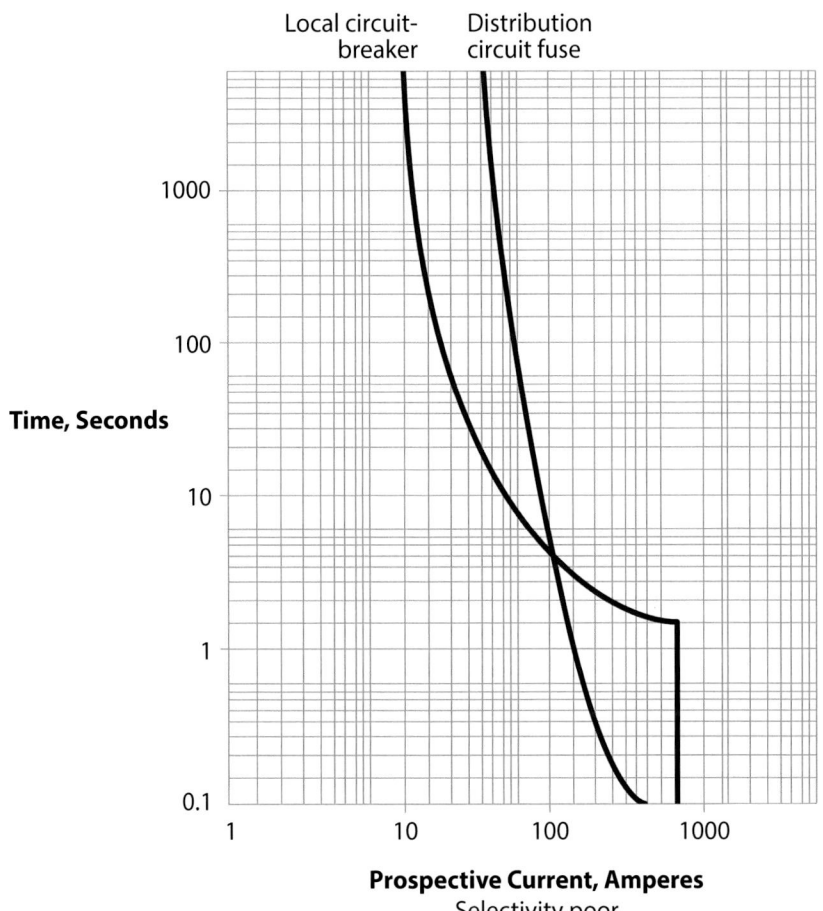

Prospective Current, Amperes
Selectivity poor

Selectivity may also need to be looked at where a circuit has more than one RCD or RCBO, regardless of their residual current settings, as the one closest to the supply may trip first.

As discussed earlier, to overcome this problem, the RCD nearest to the supply/furthest from the load should be a time-delayed or S-type RCD, meaning that it will only trip if the fault hasn't been cleared by the local RCD after a short duration in time. This time delay could be approximately 500 ms, where the local RCD should disconnect within 300 ms.

5.4 Integration of devices and components

Regulation 536.4.203 requires that all devices and components installed in an assembly, for example, a consumer unit or distribution board, are compatible with one another. The reason for this requirement being included in the IET Wiring Regulations is so that installers are aware that not all components are compatible with one another, even if they are from the same manufacturer and comply with the relevant product standards. If an installer decides to mix and match components (whether from the same manufacturer or a different one) within an assembly, they should verify with the manufacturers that all components are compatible with one another. The installer can carry out their own verification, but this will ultimately mean that the installer will become the assembly manufacturer and any problems arising from the assembly as a result of the incorrect integration of devices will be their responsibility.

Imagine this situation: you need some new brake pads for your car. At the garage, they don't have the specific ones required for your vehicle, so they use some that are very similar, meet all the product standards requirements and are of good quality – but are designed for a different make and model. You drive off and ten minutes down the road, your brakes fail and you collide with another vehicle. When the insurance company investigates, they find that the brake pads are not compatible with your vehicle. The responsibility will likely be with the mechanic who decided to use these particular pads instead of the ones that had been verified by the manufacturer.

5.5 Isolation and switching (Chapter 53 of the IET Wiring Regulations)

A typical electrical installation will contain many different switches and/or isolators. It is very important to know what the purpose of the device is, in order to select the right type. The IET Wiring Regulations require isolation and switching devices for the following reasons:

> **(a) isolation:** a means of cutting off electricity to all or part of an installation to allow safe electrical work to be carried out;

(b) **switching for mechanical maintenance:** a means of safely stopping a machine, piece of equipment or appliance from operating in any way to allow safe non-electrical work that does not involve exposure to potentially live parts;

(c) **emergency switching off:** a means of quickly stopping a machine, piece of equipment or appliance from operating in the event of an emergency;

(d) **functional switching:** a means of controlling a machine, piece of equipment or appliance, such as turning it on or off; and

(e) **undervoltage protection:** a means of stopping a machine, piece of equipment or appliance from automatically restarting following a loss of or drop in voltage, where sudden restarting could cause danger.

5.6 Isolators and switches (Chapter 46 of the IET Wiring Regulations)

Before we look at the reasons for using a switch or isolator, we need to understand the difference between them.

Isolators are rated for the current that can pass through them under normal load conditions, but not for switching ability. For example, a 100 A isolator could safely carry 100 A through it continually, but it may not be capable of switching 100 A.

If an isolator is to be used, equipment should always be powered down first. This is because arcing (when electricity sparks as it travels through the air between terminals) takes place every time a device is opened under load. The amount of arcing depends on the type of load: some loads, such as inductive loads like motors, cause bigger arcs than resistive loads, such as immersion heaters. This arcing can destroy device contacts or, as a minimum, create high resistance to the contacts, which in turn creates excessive heat through constant use.

If a device that is classed as an isolator is to be used, or positioned where it can be used, by ordinary persons as a switch (on-load), it must be de-rated in accordance with the manufacturer's instructions as a switch. For example, a 100 A isolator may only be rated to switch 40 A. The reasoning behind this is that a switch is likely to be operated far more frequently than an isolator and the rating is lower in order to reduce the wear and arcing produced at the contacts.

Isolators are often categorized in the following ways:

- **(a) SP-N (single-pole and neutral):** opens the line conductor only, but the neutral can be isolated by a link;
- **(b) double-pole:** opens the line and neutral conductors;
- **(c) triple-pole:** opens the three line conductors of a three-phase system and not the neutral;
- **(d) TP-N (triple-pole and neutral):** opens the three line conductors – the neutral can be isolated by a link; and
- **(e) four-pole:** opens all three line conductors and the neutral.

5.7 Isolation

A means of isolating all live conductors must be provided for all installations, usually in the form of a main switch, although other double-pole devices may be used. The type of installation will affect the type of isolator used.

▼ **Table 5.3** Types of installation

Installation type	Supply	Earthing arrangement	Type of device as a minimum
Dwelling and light commercial	Single-phase	▶ TT ▶ TN-S ▶ TN-C-S	▶ Double-pole switch capable of full load switching. ▶ Must be rated to switch full load current.
Commercial/ industrial	Three-phase	▶ TN-S ▶ TN-C-S	▶ TP-N device remote from the consumer unit. ▶ Triple-pole within the consumer unit.
Commercial/ industrial	Three-phase	▶ TT	▶ Four-pole.

Isolation may also be achieved for individual circuits by the protective device for the circuit.

In all situations, the device used for isolation must be capable of being secured in the 'open' or 'off' position. For main switches and circuit-breakers, this is normally achieved with a padlock and, if necessary, a special locking device. Where circuits are protected by fuses, the fuse must be capable of being fully removed and should be kept by the person who undertakes the isolation; on many sites, there may be a 'permit to work' or 'lock-off' procedure to ensure that circuits or equipment are not unintentionally energized. Devices that block off the fuseway are available and prevent other fuses from being inserted so that the circuit remains securely isolated.

More guidance on safe isolation can be found in Section 2 of this Guide.

5.8 Switching off for mechanical maintenance

Mechanical maintenance covers many tasks and relates to a variety of different items of equipment. Providing that the work does not involve exposure to electrical terminals, switching off for mechanical maintenance can be done by ordinary persons, so long as they have received the necessary instruction from their employer. Mechanical maintenance includes, amongst many other tasks:

- **(a)** changing or adjusting drive belts on a machine;
- **(b)** installing or altering the pipework to a shower;
- **(c)** cleaning air filters inside an air-handling unit (AHU); and
- **(d)** replacing lamps in luminaires.

For these situations, a device that is capable of switching full load current must be 'local' to the equipment. This means that, ideally, the device must be under the control of the person carrying out the work and they must keep the switch under their supervision, so that nobody else can switch it on.

Common devices for this purpose include:

- **(a)** fused connection units;
- **(b)** double-pole switches; and
- **(c)** plugs and socket-outlets.

If a device cannot be easily supervised, then it should be lockable (for instance, by a lockable rotary switch), ensuring that the risk of someone else switching it back on is removed. Again, there may be a 'permit to work' or 'lock-off' procedure to ensure that circuits or equipment are not unintentionally energized.

5.9 Emergency switching off

Emergency switching devices must be capable of cutting off the full load current.

In installations where items such as rotating machines can cause a danger, emergency switching is normally provided by stop buttons. When installing stop buttons to a machine or workshop, certain considerations need to be taken into account. For example, the stop button:

- **(a)** must be near to the risk, so that someone using the machine can easily reach the button if they are in danger.
- **(b)** must not be capable of being reset from another place unless the reset is by a key switch.
- **(c)** should latch in the 'off' position where (b) is not possible. This ensures that the person resetting the circuit must be where the danger happened, so that they can see that the danger has been removed before resetting the circuit.
- **(d)** should be key-operated where the stop button could be reset by untrained persons, as, for example, with a stop button in a school workshop.

5.10 Functional switching

This is probably the most common type of switch found in an installation. Functional switching allows equipment to be switched on or off as needed, and should be in a convenient place, where the switch is operated manually. An example would be a light switch for a room, located by the entrance door to the room.

Where a room has more than one entrance, two-way and/or intermediate switching would normally be provided.

Functional switches may also be in the form of:

- **(a)** time clocks;
- **(b)** passive infra-red (PIR) sensors;
- **(c)** plugs and socket-outlets (rated at or below 32 A);
- **(d)** switches on socket-outlets;
- **(e)** contactors;

(f) standard light switches;
(g) push switches; and
(h) switched fused connection units.

Devices that should never be used as a functional switch include:

(a) fuses;
(b) luminaire connections, such as plug-in ceiling roses;
(c) unswitched fused connection units; and
(d) socket-outlets rated above 32 A.

Where a functional switch is controlling discharge luminaires, such as fluorescent fittings, the switch must be rated high enough to manage the surge current that happens when the load is first switched on. Typically, a 6 A lighting circuit with fluorescent lights will have 10 A switches for this reason.

5.11 Undervoltage protection

Undervoltage protection is required where the unexpected restarting of a machine could cause danger.

Imagine a grinding wheel that is being used when a power cut occurs. If the grinding wheel does not have undervoltage protection, the wheel will dangerously and unexpectedly start spinning again when the power comes back on. This can be a huge risk to anyone near to or touching the wheel.

A way to overcome this problem is to install a contactor with start and stop buttons next to the machine.

With this type of arrangement, the machine starts by pressing the start button. The contactor coil is energized, which, in turn, closes the contactor switch giving power to the machine. If the power fails, the contactor coil drops out, opening the circuit. If the power then comes back on, there is no danger, as the contactor requires somebody to push the start button for the machine to come back on.

▼ **Figure 5.11** A typical motor circuit

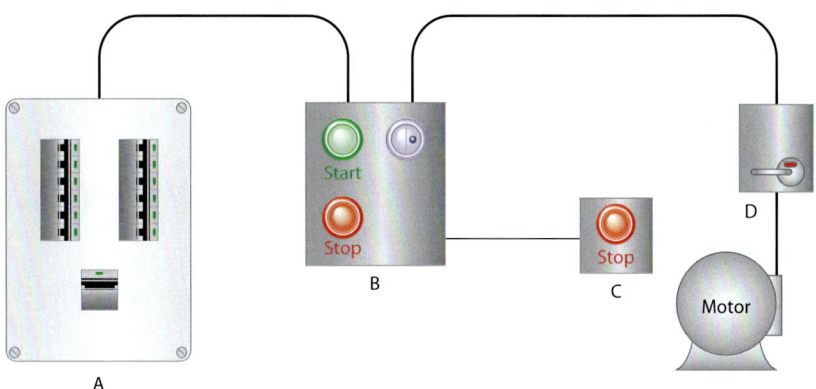

The diagram above shows a typical motor circuit, for example, one used to drive a mechanical pump. Each device labelled A to D performs the following function:

Device	Description	Function
A	Distribution board	▶ Full isolation (main switch) ▶ Circuit isolation (circuit-breaker) ▶ Protection (overcurrent)
B	Start/stop contactor with switch	▶ Functional switching ▶ Partial isolation ▶ Emergency switching ▶ Undervoltage protection ▶ Overload protection
C	Remote stop button	▶ Emergency switching
D	Local switch	▶ Switching for mechanical maintenance ▶ Isolation (so long as it is lockable)

5

	Test your knowledge
1.	What is an overload, as defined in Part 2 of the IET Wiring Regulations?
2.	Name a hazardous material that may be found in the carriers of BS 3036 fuses.
3.	What does the I_{cs} rating of a circuit-breaker relate to?
4.	Determine, using the graphs in Appendix 3 of the IET Wiring Regulations, the approximate time it takes for each of the following devices to disconnect with an overload of 60 A: (a) 32 A BS 88-3; (b) 32 A BS 88-2; (c) 30 A BS 3036; and (d) 32 A BS EN 60898, Type B.
5.	What is the maximum residual current rating ($I_{\Delta n}$) permitted for an RCD that provides additional protection in accordance with the IET Wiring Regulations?
6.	Which chapter from Part 5 of the IET Wiring Regulations covers selectivity between RCDs?
7.	What specific requirements within Chapter 53 of the IET Wiring Regulations should be met where an 'off-load' device for isolation is used within an installation?
8.	List three examples of common functional switches.
9.	What danger may exist if a machine doesn't have suitable undervoltage protection?
10.	What would commonly be used to provide undervoltage protection for a machine?

Earthing and Bonding   6

This Section provides information on the following topics:
- ▶ **What earthing is and why we need it**
- ▶ **Earthing arrangements**
- ▶ **What bonding is and why we need it**
- ▶ **Earth fault loop impedance (EFLI)**

To understand how earthing works, you must first understand that electricity will take the easiest and quickest path every time. When it leaves one side of a transformer, it wants to get to the other side as quickly and as easily as possible (the path of least resistance).

It works in the same way as if you wanted to drive a car from one side of a city to another and had the choice of driving through the congested city centre or taking the ring road (bypass). It goes without saying that you would take the quickest route (the path of least resistance). Electricity is the same – well, sort of!

6.1 What is earthing and why do we need it?

The The Electricity Safety, Quality and Continuity Regulations 2002 (ESQCR) state that the electricity distributor is, in most circumstances, obliged to provide the consumer with an earthing facility for a new supply to a low voltage (LV) installation. There may be safety reasons that prevent the distributor from providing an earth: for example, the neutral conductor may be susceptible to damage when used as a TN-C-S arrangement. An example of this would be an overhead supply to a house in the countryside. If the neutral conductor were to be severed by a farmer in a tractor, 230 V could appear between true Earth and all metalwork within the installation.

The distributor may then refuse to provide a means of earthing in this case, and the consumer must install a TT earthing arrangement instead. Ultimately, the consumer is responsible for ensuring that the installation is suitably earthed.

The purpose of earthing is to provide a path for current to flow through (in the event of a line-to-earth fault), so that the protective device operates within the required time (see Section 5 of this Guide for more information on how protective devices work). For this to happen, the current needs to flow back to the neutral terminal of the distribution transformer (the origin), creating a complete circuit, albeit a completed fault circuit. To achieve this, there needs to be a good conductive path from wherever the fault occurs, back to neutral at the transformer, via the main earthing terminal (MET) at the consumer installation. The various ways in which this is achieved is called earthing.

6.1.1 Why do we call it earthing?

We call it earthing simply because in most electrical systems in the UK and many other countries, the neutral at the distribution transformer is connected to the general mass of Earth. By 'Earth' (note capital E!), we mean the thin layer of material that covers our planet, whether it be soil, mud, rock, clay or anything else you might find below your feet, whereas all the green-and-yellow conductors are connected to the earth (lower case 'e'!) of the electrical installation. That's right! There is a physical connection, by means of an earth rod and, in most cases, a copper mesh, between neutral and Earth in almost all distribution transformers in the UK.

When you remove the cover to the electrical terminals of most household appliances, you will see that there is a cable identified with green-and-yellow insulation, along with a cable with brown insulation and a cable with blue insulation. The green-and-yellow cable is the circuit protective conductor (cpc) and connects the exposed-conductive-parts of appliances and the installation to earth. A cpc will carry fault currents to the consumer unit, where the earthing conductor will send them on their way to Earth!

Chapter 54 of the IET Wiring Regulations provides the requirements for earthing arrangements and protective conductors.

> **Example: a light fitting**
>
> A light fitting has a metal body (exposed-conductive-part), which, in normal circumstances, is not live. But what if, when being moved around or accidentally knocked, the cable becomes damaged and the live conductor makes contact with the metal casing?
>
> Remember that electricity takes the easy route. In the case of this light fitting, imagine that the filament of the lamp is a congested city: it is difficult for the electricity to pass through. Meanwhile, the metal casing is like a clear motorway – a nice big chunk of conductive material to pass through.

When designing an electrical installation, the type of electrical supply from the distribution network operator (DNO) must be taken into consideration. This is especially important on a TT system, where the general mass of the Earth is relied upon for fault current to flow around the EFLI path. It is common for this type of system to rely on residual current device (RCD) protection, due to the high impedance of the EFLI paths. The following figures show how the cpc and earthing conductor provide a path for fault current to flow to the supply earthed neutral.

6.1.2 What is a fault to earth?

By this stage, you will know that simple single-phase circuits consist of two live conductors: line and neutral. When one of these conductors, or part of some equipment that carries current, becomes damaged and connects with an exposed-conductive-part (see Section 1), it creates a path for the current to flow to the cpc. This cpc is connected between the exposed-conductive-parts and the MET of that installation. If this offers the path of least resistance, then most of the current flow will divert and travel through the cpc back to the earthed neutral of the distribution transformer (source). As long as the installation meets the requirements of the IET Wiring Regulations, there should be sufficient current to flow to operate the protective device within the required time.

6.1.3 What happens if there is a bad connection to earth or no connection at all?

Keep in mind that installations in the UK have the neutral conductor connected to the general mass of the Earth at the distribution transformer and that the current will travel through the general mass of the Earth to get back to the transformer if this is the easiest path.

Now imagine an installation that is not correctly earthed, such as a corroded protective conductor (creating higher resistance) or an installation that is not earthed at all (no path). If the same fault was to occur and the exposed metallic part of the equipment became live, it would stay live until something or someone created a path for the current to flow to earth or through the general mass of the Earth. This might happen if a person was to touch a metal part that had become live, as a result of a fault, while standing on the ground (Earth) or touching a piece of conductive material that emerged from the ground, such as metallic pipework or structural steel (for more information about extraneous-conductive-parts, see Section 1). The consequences have proven fatal on many occasions in the past.

6.1.4 Earthing arrangements

Terre neutral-connected-separated (TN-C-S) arrangement

For more information about terre neutral-connected-separated (TN-C-S) arrangements, see Regulation 542.1.2.2 in the IET Wiring Regulations.

At the distribution transformer, the neutral conductor is connected to Earth by means of an earth electrode. The neutral is now 'earthed' and the term 'terre', which is French for Earth, describes that the source is earthed. From the transformer, a single conductor performs both the neutral and earthing functions and is known as the protective earthed-neutral (PEN conductor). Together with the line conductor, the PEN conductor forms part of a composite cable that is buried in the ground and run to the consumer's property. At this point, the arrangement is TN-C, with the C indicating that the earthing and neutral functions are combined in the single conductor. When the cable enters the property, a special distribution fuse carrier has an extra terminal, often referred to as the protective multiple earth (PME) terminal, which provides an earthing connection for the electrical installation.

You can see from Figure 6.1 that the TN-C-S arrangement uses the neutral supply conductor (PEN) as the means of earthing for the installation. This means that if a line-to-earth fault were to occur within this installation, for example, due to a faulty kettle, the fault current would travel through the cpc to the MET of the installation.

From there, the installation's earthing conductor, usually a 16 mm² green-and-yellow insulated copper conductor, would carry the fault current to the PME earthing terminal. From there, the fault current would travel through the combined PEN conductor (the blue one), back to the transformer. This flow of current would be sufficient (if the installation complied with the IET Wiring Regulations) to operate the protective device in the required time.

▼ **Figure 6.1** TN-C-S arrangement

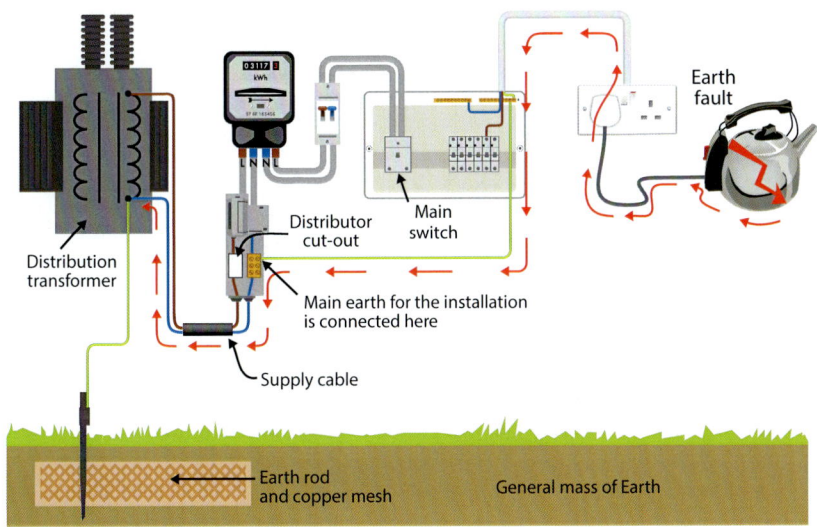

Terre Neutral-Separated (TN-S) arrangement

For more information about terre neutral-separated (TN-S) arrangements, see Regulation 542.1.2.1.

Terre is connected to the neutral conductor at the distribution transformer, as it is with the TN-C-S arrangement. However, in the TN-S arrangement, the earth conductor from the transformer to the property is separate. The separate conductor is usually the steel wire armour (SWA) on the supply cable, which in most cases, provides adequately low resistance.

You can see from Figure 6.2 that the TN-S arrangement uses a separate conductor as a means of earthing the installation. This means that if a line-to-earth fault were to occur on this installation, for example, due to a faulty kettle, the fault current would travel through the cpc to the MET of the installation down to where the supply cable enters the property. After this, it would travel through a separate protective earth conductor back to the transformer. This flow of current would be sufficient (if the installation complied with the IET Wiring Regulations) to operate the protective device in the required time.

▼ **Figure 6.2** TN-S arrangement

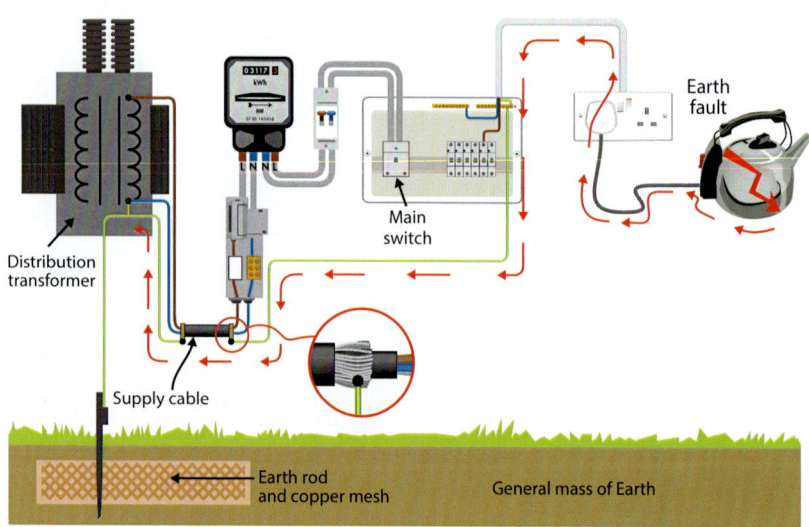

Terre terre (TT) or earth earth arrangement

For more information about terre terre (TT) or earth earth arrangements, see Regulation 542.1.2.3.

At the distribution transformer, the neutral conductor is connected to Earth by means of an earth electrode. The neutral is now 'earthed' and the term 'Terre', which is French for Earth, describes that the source is earthed. This connection of the neutral conductor to earth at the distribution transformer is as it is with the TN-C-S and the TN-S arrangements. However, with a TT earthing arrangement, the earth conductor from the transformer to the property is actually the general mass of the Earth! As a result, to create a path for current to flow from a consumer's installation to the transformer, an earth electrode is driven into the ground close to the consumer's installation. This provides a good return fault path via which the current can flow to the general mass of the Earth. From here, it will flow through the ground to the distribution transformer.

However, the general mass of Earth is not the most effective conductor and varies significantly depending on location and moisture content in the ground.

You can see from Figure 6.3 that the TT arrangement uses the general mass of Earth as a means of earthing the installation. This means that if a line-to-earth fault were to occur on this installation (for example, due to our faulty kettle), the fault current would travel through the cpc to the MET of the installation, along the earthing conductor, down to the earth rod and then through the general mass of the Earth back to the distribution transformer. This flow of current would be sufficient (if the installation complied with the IET Wiring Regulations) to operate an RCD in the required time.

▼ **Figure 6.3** TT arrangement

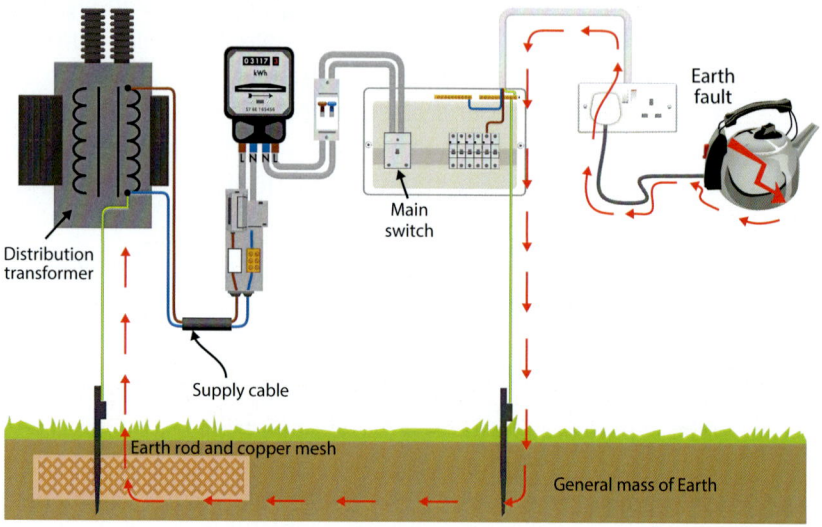

6.2 Protective equipotential bonding

Section 544 of the IET Wiring Regulations covers the requirements for protective equipotential bonding conductors, including information relating to:

(a) the required cross-sectional area (csa) of the main protective bonding conductors and supplementary bonding conductors;
(b) where and how they should be connected in relation to the consumer unit; and
(c) where metallic pipework enters the building, etc.

Section 4 of the *On-Site Guide* provides comprehensive guidance on the installation of bonding conductors.

6.2.1 What is equipotential bonding and why do we need it?

In an installation where the protective measure is automatic disconnection of supply (ADS), where the protective device, e.g. the RCD, detects a fault and disconnects the supply to the circuit automatically, protective equipotential bonding must be provided in accordance with the IET Wiring Regulations. The purpose of main protective equipotential bonding is to reduce touch voltages during an earth fault that may exist between:

> **(a)** extraneous-conductive-parts (metallic parts that do not form part of the installation, but may introduce Earth potential, such as structural steel and metallic pipework) and other extraneous-conductive-parts;
> **(b)** extraneous-conductive-parts and exposed-conductive-parts; and
> **(c)** exposed-conductive-parts and Earth.

6.2.2 Potential difference

One thing to remember is that the general mass of the Earth is considered to have zero potential or zero volts. This means that the difference between, let's say, a single-phase line conductor and the general mass of the Earth will be 230 V, giving a potential difference of 230 V. If something were to create a path between the two, it would have the full 230 V placed across it, whether that something was a piece of conductive material or a human body.

Something you should never try

As a child, I often wondered why birds could fly on and off bare electrical transmission cables without being electrocuted. People would generally explain to me that it was because they were not touching two wires at the same time. While this is true to some extent, it did not become properly clear to me until I had a better understanding of how electricity actually works. Whether it's a bird – or even me – hanging onto an overhead cable, the principles are the same. If I were to hang onto one part of the cable with my right hand and another part of the same cable with my left hand, I would be able to hang there quite happily (albeit 20 or so metres above the ground). This is because the potential difference between the two parts of the cable in my hands would be zero.

This can be proven with a voltmeter by holding the two probes onto the same live conductor a short distance apart (it is recommended that this is done under the supervision of a lecturer and on an extra-low voltage system). The value shown will be zero, or at least, very close to zero. This is, essentially, what we are trying to achieve when we install equipotential bonding: to achieve as close to zero potential as possible between conductive parts.

Hopefully this helps, when trying to understand 'equipotential'. It essentially means 'equal potential': i.e., to have the same potential at two or more points. Although this is almost impossible to achieve, any reduction in the potential difference between two parts is an improvement to the safety of an installation during an earth fault.

6.2.3 What happens during a fault?

During a fault to earth, the current will flow around the EFLI path until it reaches a magnitude sufficient to cause automatic operation of the protective device. However, if something or someone were to come into contact with a live exposed-conductive-part that wasn't earthed and an extraneous-conductive-part at the same time, and the installation had insufficient protective equipotential bonding, they could immediately create a path for current to flow through the body between these two parts. The voltage between these two parts is known as 'touch voltage' (U_t).

Touch voltage can give an electric shock, which can cause serious injury or even death. Equipotential bonding, in accordance with the IET Wiring Regulations, can reduce touch voltage during an earth fault to an acceptable level (low enough to prevent a serious risk from electric shock), until the protective device operates.

6.2.4 How does protective equipotential bonding work?

Remember that electricity will always take the easiest path back to the neutral at the supply transformer. If the easiest path is through the cpc and the protective equipotential bonding conductors, then that's where most of the current will flow. If there is an alternative route, such as through a person to an extraneous-conductive-part, then that person is going to receive a nasty shock. This is why it is important not only to provide protective equipotential bonding, but also to ensure that the csa of the conductor is sufficient to provide an easy path through which current can flow.

In the figures below, a line-to-earth fault has occurred on a TN-C-S system and has introduced a voltage potential between the exposed-conductive-part of the electrical equipment and the extraneous-conductive-part (metal pipework). The earth fault path (earth fault loop) is shown by the purple dotted line.

In Figure 6.4 there is no protective equipotential bonding. When an earth fault occurs, the touch voltage may be fairly high between the exposed-conductive-part (the metallic casing of the kettle) and the extraneous-conductive-part (the metallic pipework rising from the ground). This is because the man in the diagram is creating a relatively easy path for the electricity to flow back to the supply transformer via the metal pipework and the general mass of the Earth. In this example, there is a touch voltage of 143 V, which is travelling through the person. This has the potential to cause serious injury or even death.

▼ **Figure 6.4** Installation without protective equipotential bonding

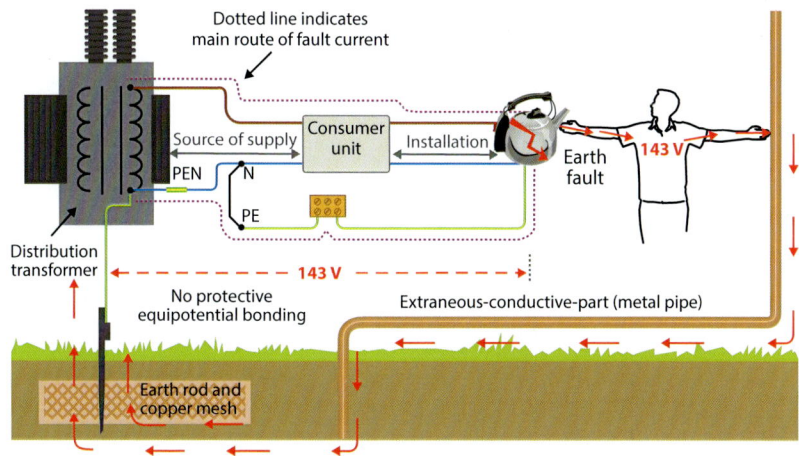

In Figure 6.5 the installation has protective equipotential bonding, which connects the extraneous-conductive-part as it enters the property to the MET at the consumer unit. By installing this additional conductor, the potential difference that exists during a fault to earth is reduced. This will significantly lessen the magnitude of the touch voltage that can exist between the exposed-conductive-part (the metalwork of the lamp) and the extraneous-conductive-part of the installation.

In Figure 6.5, we can see that the touch voltage has been reduced to 50 V.

▼ **Figure 6.5** Installation with protective equipotential bonding

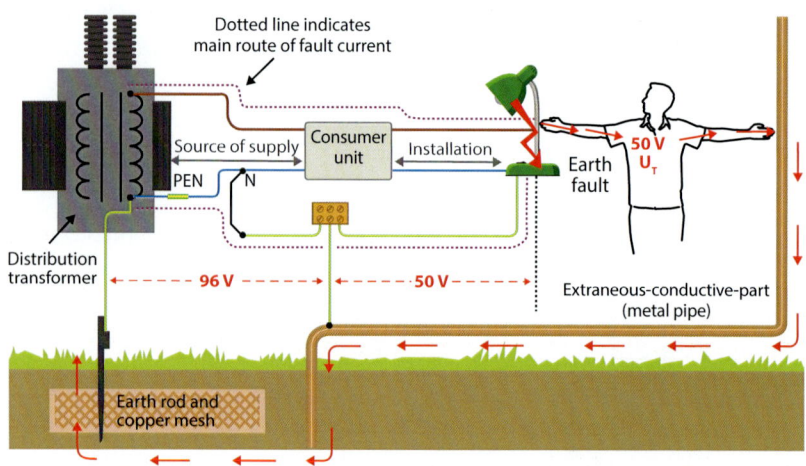

6.2.5 Connecting two or more extraneous-conductive-parts using a single conductor

If a single protective bonding conductor is used to bond two or more extraneous-conductive-parts, it should be one continuous length of cable, so that in the event of any part of the bonding group becoming detached from the extraneous-conductive-part, there will still be continuity between any other extraneous-conductive-parts and the MET.

Guidance on the correct size of bonding conductors to be used can be found in Table 4.4(i) and 4.4(ii) of the *On-Site Guide*. The minimum size that should be used in any installation is 6 mm². The size of the main protective bonding conductor is directly related to the size of the incoming live conductors.

In Figure 6.6 an example of how the conductor should be connected to the terminal (unbroken) is shown.

▼ **Figure 6.6** Unbroken bonding conductor

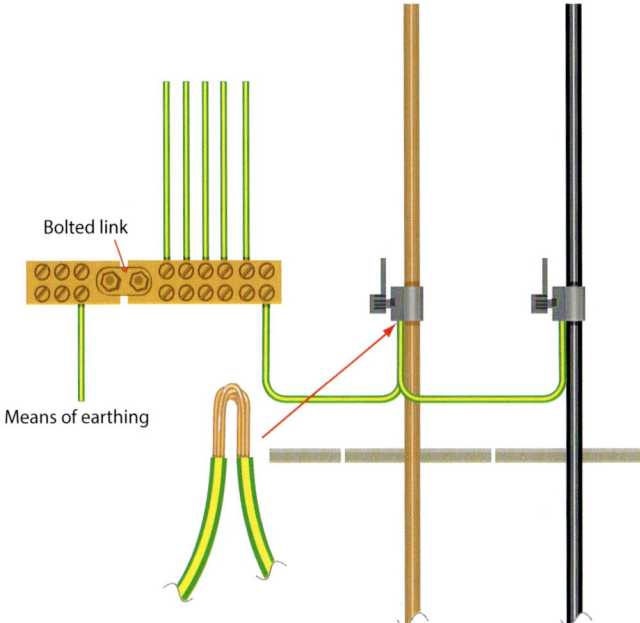

6.3 Supplementary equipotential bonding

The principles of supplementary equipotential bonding are the same as equipotential bonding. The only difference is that it further reduces any potential voltage in a particular part of an installation during an earth fault that may arise between exposed-conductive-parts and extraneous-conductive-parts.

There are certain circumstances where it is required by the IET Wiring Regulations to have supplementary equipotential bonding, particularly in installations and locations where there is an increased risk of electric shock, such as locations containing swimming pools. Part 7 of the IET Wiring Regulations provides more information on the locations that require this additional protection.

Section 4 of the *On-Site Guide* provides information on how to install protective equipotential bonding. Table 4.6 is particularly useful for identifying the required csa of the bonding conductors to achieve the level of protection required.

All the requirements and guidance available in the IET Wiring Regulations and associated guidance for protective equipotential bonding are intended to help you reduce the voltage between parts that may exist during an earth fault and, ultimately, reduce the severity of electric shock that someone may experience.

	Test your knowledge
1.	Is the earth conductor of an electrical installation considered to be a live conductor?
2.	If the distributor is unable to provide a connection to earth at the consumer's installation, what type of earthing arrangement would be considered?
3.	The purpose of earthing is to provide a path for current to flow in the event of a fault so that the protective device will operate. To which terminal at the distribution transformer would the earthing conductor be connected?
4.	What is used to connect the neutral side of the transformer to the general mass of the Earth?
5.	Write the full description of the following abbreviations: (a) TN-S; (b) TN-C-S; and (c) TT.
6.	Is the metal casing of a Class 1 light fitting an extraneous-conductive-part or an exposed-conductive-part?
7.	Is a structural reinforced steel girder rising from the ground an exposed-conductive-part or an extraneous-conductive-part?
8.	What event would result in a touch voltage appearing within an electrical installation?
9.	What would be the expected reading if the two probes of a voltmeter are connected to the same energized line conductor?
10.	Why is it required to have one continuous conductor for protective equipotential bonding when it is connected to more than one extraneous-conductive-part?

Cable Calculations 7

This Section provides information on the following topics:
- ▶ Why we need cable calculations
- ▶ Current rating of equipment and protective devices and current-carrying capacity of cables
- ▶ Rating factors
- ▶ Voltage drop

7.1 Why do we need cable calculations?

When a circuit is in use, there is current travelling through the conductors. The more electrical appliances that are in use and the higher the power requirements of the appliances, the more current will flow through the cable.

Cables have a limit! If the installation hasn't been designed correctly and the cable for the circuit is too small (i.e. it does not have the required current-carrying capacity), there is a high risk of the cable overheating and causing a fire. This could ultimately lead to the destruction of property and present a serious risk to life.

It is therefore necessary to consider what the circuit is going to be used for, the environment in which the cable will be installed and the type of cable to be used. As long as the installation meets the requirements of the IET Wiring Regulations, the cable should be adequate to carry the design current under the specified installation conditions.

7.2 Current rating of equipment and protective devices and current-carrying capacity of cables

7.2.1 Current ratings and capacities

There are various symbols and abbreviations that you need to be familiar with to efficiently calculate the required size of a cable for an installation. Below, we discuss the different types of current ratings. You'll notice that each begins with 'I', which is the SI symbol for current.

I_b (design current)

This is the current that will be used by an item of electrical equipment under normal operating conditions. For example, if you have a **1,000 W** heater supplied at **230 V**, I_b will be the maximum current that will flow to operate the appliance. In this example, the design current will be **4.35 A** (see later in this section for an explanation of how to work this out). For circuits designed to supply multiple appliances, such as cooker circuits, diversity should be taken into consideration (guidance on this is given in Section 12).

I_n (rating of protective device)

This is the rating of the device that is designed to operate and break a circuit should a fault or overload occur. It is the current that the device can carry for an indefinite period without being overloaded.

I_t (tabulated current-carrying capacity of cable)

This is the current-carrying capacity of the cable without the application of any correction factors or installation conditions taken into consideration. Essentially, it is the maximum current a cable will be able to carry (see Appendix 4, Tables 4D1A to 4D5 of the IET Wiring Regulations).

I_z (current-carrying capacity of a cable for continuous service)

This is the current-carrying capacity of the cable for continuous service with all necessary correction factors applied and with installation conditions taken into account. For example, if a cable has a long distance to travel and it has to pass through thermal insulation, factors need to be applied and its current-carrying capacity will be lower than that of the specified value in Appendix 4, Tables 4D1A to 4D5 of the IET Wiring Regulations.

7.2.2 Rating factors to consider in calculations

What are correction factors?

The resistance of a cable is affected by heat. As the cable gets warmer, the resistance increases. The retention of heat in a cable is affected by many factors. Listed below are some of the factors that should be considered when designing an installation.

C_a (ambient temperature)

This is the temperature of the environment in which the cable will be installed.

▼ **Figure 7.1** Temperature of environment

C_c (cables buried in the ground)

When cables are buried in the ground, the heat generated is not able to readily dissipate and the temperature of the conductors will rise, causing the resistance of the conductors to increase when the circuit is in use.

▼ **Figure 7.2** Cables buried in the ground

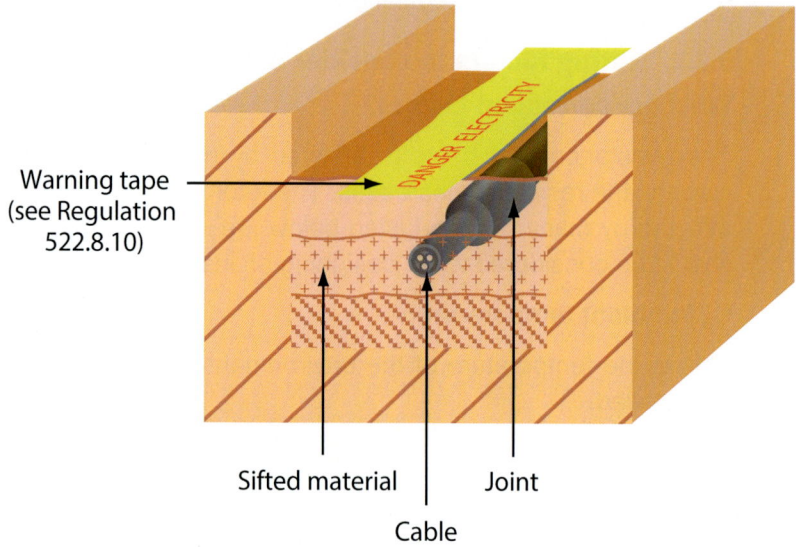

C_d (depth of burial)

Heat dissipation will vary depending on the depth at which the cable is buried.

▼ **Figure 7.3** Depth of cables buried in the ground

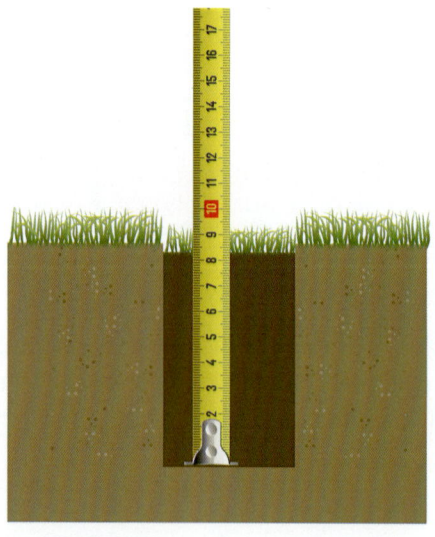

C_f (semi-enclosed fuses – see BS 3036)

Where a semi-enclosed fuse to BS 3036 is used to protect a cable against overload, the cable needs to be derated by 0.725 in addition to any other necessary factor; this is one of the disadvantages of using this type of protective device.

▼ **Figure 7.4** BS 3036 fuses

C_g (grouping)

Various influences affect cables when they are grouped together. The build-up of heat has the most impact on the resistance of conductors.

▼ **Figure 7.5** Cable grouping

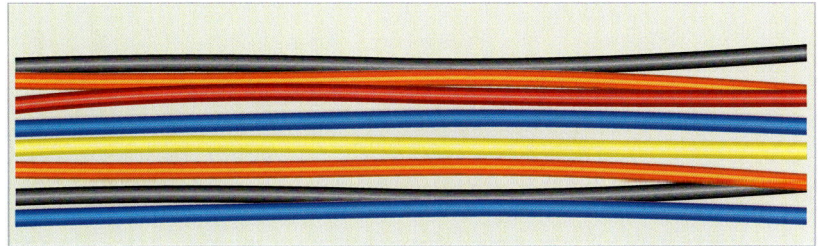

C_i (thermal insulation)

Thermal insulation, common in most installations, can cause cables to overheat when installed in or alongside it.

▼ **Figure 7.6** Cable run through thermal insulation

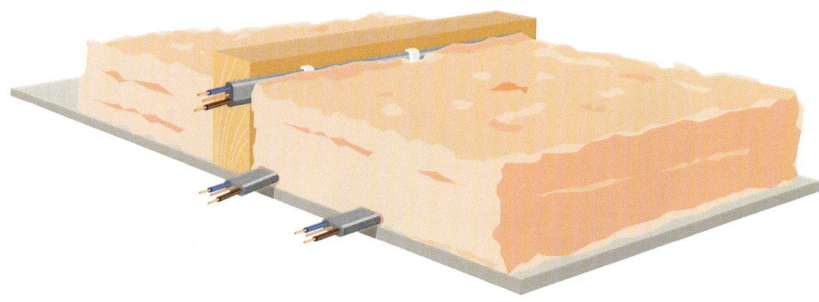

C_s (soil resistivity)

The resistivity (resistance) of soil can vary, depending on the location, moisture content, etc.

▼ **Figure 7.7** Cable run underground

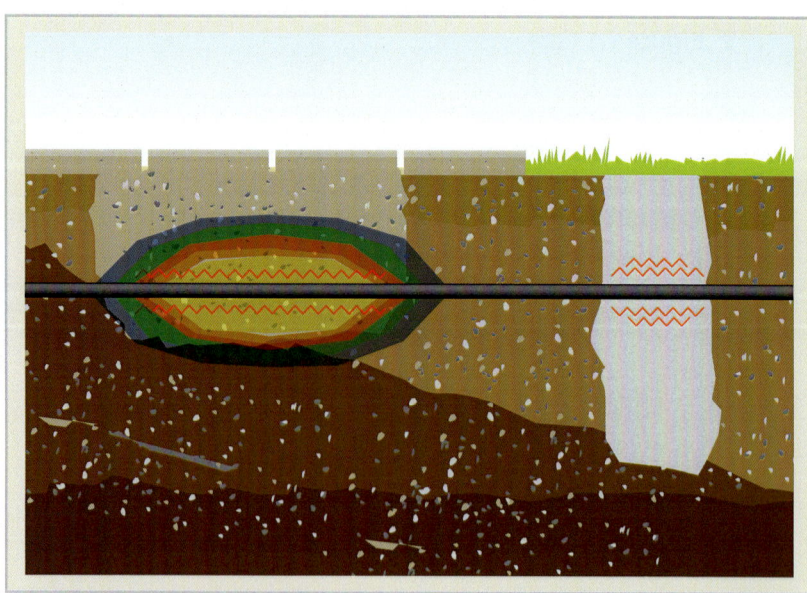

> **Remember...**
> To help you remember these, try to associate the letters with the correction factors. For example:
> - C_a = Rating factor **A**mbient temperature (see Table F1 of the *On-Site Guide*)
> - C_c = Rating factor Buried (think of it as **C**overed)
> - C_d = Rating factor **D**epth of burial
> - C_f = Rating factor **F**use (BS 3036 only)
> - C_g = Rating factor **G**rouping (see Table F3 of the *On-Site Guide*)
> - C_i = Thermal **I**nsulation (see Table F2 of the *On-Site Guide*. Note that for cables installed using the methods described in Tables F4(i), F5(i) and F6, the rating factor that needs to be used is 1, as the allowance for thermal effects has already been taken into account)
> - C_s = **S**oil resistivity

7.3 How design current, protective device rating and current-carrying capacity relate to each other

7.3.1 Design current, protective device rating and cable rating

I_b (design current): should always be equal to or less than I_n.

I_n (overcurrent rating of protective device): should always be equal to or less than I_t.

I_t (tabulated current-carrying capacity of a cable): will invariably be higher than I_z.

I_z (current-carrying capacity of cable for continuous use under particular installation conditions): providing the value is equal to or greater than the rating of the protective device (I_n), the cable will be suitably protected.

This can be viewed as:
- $I_b \leq I_n \leq I_t$
- $I_n \leq I_z$

If I_b was higher than the rating of the protective device, it would cause the device to operate whenever the circuit was in use and would, therefore, be impractical.

If the protective device was rated higher than I_z, there would be a risk of the cable overheating and catching fire before the protective device could operate.

7.3.2 Identifying the values to be used when calculating the required rating of cable and protective device

Identifying the supply voltage

In most UK dwellings, the supply voltage will be 230 V single-phase. Note that the supply voltage at the terminals of an item of equipment is very rarely 230 V when we measure it. Many factors affect the supply voltage, such as the length of run of the cables or electrical loading in the area. The Electricity Safety, Quality and Continuity Regulations 2002 (ESQCR) permit the supply voltage to fluctuate between −6 % and +10 %, which means that when we measure the voltage at the supply terminals on an electrical installation, we would expect the voltage to be between 216.2 V or 253 V. With this in mind, we base all calculations on a nominal voltage of 230 V. The voltage and number of lines that are needed to operate a particular piece of equipment must be determined at the design stage of the installation.

Identifying the power rating of the appliance or equipment

Most appliances or equipment will have voltage and power ratings clearly identified on them and these should also be stated in the manufacturer's instructions.

Once both the voltage and the power ratings have been identified, you can work out the design current to determine the necessary size of cable, using the following formula.

Working out the design current

For example, if an installation consists of one storage heater with a power rating of **3 kW**, that is what will be used in the calculation. If there are three storage heaters at **3 kW** each, the total that will be used in the calculation will be **3 × 3 kW (9 kW)**.

$$I_b(\text{Design I}) = \frac{P(\text{Power W})}{V(\text{Volts})}$$

In this example, the equation would look like this:

$$I_b(\text{Design I}) = \frac{9000 (9 \text{ kW})}{230} = 39.1 \text{A}$$

Determine the appropriate protective device rating

Once the design current has been calculated, the appropriate protective device rating (I_n) can be selected, which is always equal to or greater than I_b.

Determine the size of the cable

By determining I_t using the tables provided in Appendix 4 of the Wiring Regulations, a cable with sufficient current-carrying capacity can be found. There are many different installation methods that help determine the required cross-sectional area (csa) of the conductor. A description of each method can be found at the top of each table.

Once I_n has been selected, the requirements for I_t can be identified, using the formula:

$$I_t \geq \frac{I_n}{C_a \times C_c \times C_d \times C_f \times C_g \times C_i \times C_s}$$

This is assuming that there are seven factors to take into consideration. If, for example, you have no thermal insulation to run cables through, you will simply remove C_i from the calculation.

If the only factor to take into consideration is the ambient temperature C_a, it will be the only factor in the calculation, as shown below:

$$I_t \geq \frac{I_n}{C_a}$$

7

Example

A circuit is to be installed to supply a 230 V appliance with a power rating of 5 kW. The ambient temperature in the room where it is to be installed is 35 °C. The 70 °C thermoplastic flat sheathed cable will need to be run through 2.5 m of thermal insulation exceeding 100 mm in thickness in a stud wall, touching the inner wall. The total length of the cable run is 10 m.

The calculation is as follows:

To determine the design current (I_b):

$$I_b = \frac{P}{V}$$

P = 5 kW (5,000 W)

V = 230 V

$$I_b = \frac{5,000\,W}{230\,(V)} = 21.73\,A$$

Now that the design current has been calculated, the other factors can be used to work out the size of the cable required. We already have one value:

I_b = 21.73 A

As previously stated, the protective device (I_n) must be rated either equal to or higher than the design current (I_b), so it should be the lowest rating available:

21.73 A ≤ I_n

This would mean using a protective device with a rating of 21.73 A or higher. For this example, a circuit-breaker to BS EN 60898 with a rating of **32 A** is used.

The first value we have is: **I_n = 32 A**

The next value that can be used is the correction factor for the ambient temperature (C_a). This can be obtained using Table 4B1 from Appendix 4 of the IET Wiring Regulations and Table F1 from the *On-Site Guide*. The cable being used is flat 70 °C (the maximum operating temperature of the cable) thermoplastic and the ambient temperature is 35 °C, so the factor to be used is 0.94.

▼ **Figure 7.8** Checking Table 4B1 from Appendix 4 of the IET Wiring Regulations

TABLE 4B1 – Rating factors (C_a) for ambient air temperatures other than 30 °C

Ambient temperature[a] °C	Insulation				
	60 °C thermosetting	70 °C thermoplastic	90 °C thermosetting	Mineral[a]	
				Thermoplastic covered or bare and exposed to touch 70 °C	Bare and not exposed to touch 105 °C
25	1.04	1.03	1.02	1.07	1.04
30	1.00	1.00	1.00	1.00	1.00
35	0.91	$C_a = 0.94$	0.96	0.94	0.96
40	0.82	0.87	0.91	0.85	0.92
45	0.71	0.79	0.87	0.78	0.88
50	0.58	0.71	0.82	0.67	0.84
55	0.41	0.61	0.76	0.57	0.80
60		0.50	0.71	0.45	0.75
65		–	0.65	–	0.70
70		–	0.58	–	0.65
75		–	0.50	–	0.60
80		–	0.41	–	0.54
85		–	–	–	0.47
90		–	–	–	0.40
95		–	–	–	0.32

a For higher ambient temperatures, consult manufacturer.

As we can see from Table 4B1, $C_a = 0.94$.

The next value to be calculated is the thermal insulation factor C_i. This can be obtained using Table 4D5 from the IET Wiring Regulations or Table F6 from the *On-Site Guide*, where the current-carrying capacity of the cable can be found. This table shows the various different installation methods and how they affect the current-carrying capacity of the cable. As described earlier in this section, if the installation method can be found in this table, the rating factor to be used in the calculation is 1.

7

Now that all possible values and factors have been obtained, the calculation for I_t is:

$$I_t \geq \frac{I_n}{C_a}$$

$$I_t \geq \frac{32}{0.94}$$

$$I_t = 34.04$$

Using the value **34.04 A**, refer to Table 4D5 to identify the size of the cable to be used. The installation method, as mentioned earlier, is 102.

TABLE 4D5 – 70 °C thermoplastic insulated and sheathed flat cable with protective conductor (COPPER CONDUCTORS)

CURRENT-CARRYING CAPACITY (amperes) and VOLTAGE DROP (per ampere per metre):

Ambient temperature: 30 °C
Conductor operating temperature: 70 °C

Conductor cross-sectional area	Method 100# (above a plasterboard ceiling covered by thermal insulation not exceeding 100 mm in thickness)	Method 101# (above a plasterboard ceiling covered by thermal insulation exceeding 100 mm in thickness)	Method 102# (in a stud wall with thermal insulation with cable touching the inner wall surface)	Method 103# (in a stud wall with thermal insulation with cable not touching the inner wall surface)	Reference Method A* (enclosed in conduit in an insulated wall)	Reference Method B* (enclosed in conduit on a wall or in trunking etc.)	Reference Method C* (clipped direct)	Voltage drop (per ampere per metre)
1	2	3	4	5	6	7	8	9
(mm²)	(A)	(A)	(A)	(A)	(A)	(A)	(A)	(mV/A/m)
1	13	10.5	13	8	11.5	13	16	44
1.5	16	13	16	10	14.5	16.5	20	29
2.5	21	17	21	13.5	20	23	27	18
4	27	22	27	18.5	26	30	37	11
6	34	27	35	23.5	32	38	47	7.3
10	45	36	47	32	44	52	64	4.4
16	57	46	63	42.5	57	69	85	2.8

You can see that, based on the information provided in this section, and using the table above, a 6 mm² cable can be used for this installation, providing the voltage drop is within the required percentage.

7.4 Voltage drop

7.4.1 What is voltage drop?

Regulation 525 of the Wiring Regulations outlines the requirements for voltage drop and Table 4Ab from Appendix 4 provides the maximum permissible voltage drop for consumer installations.

When current travels through cable, energy is being used.

The following scenario can be related to the energy used when electricity is travelling through a conductor (the 'tunnel analogy').

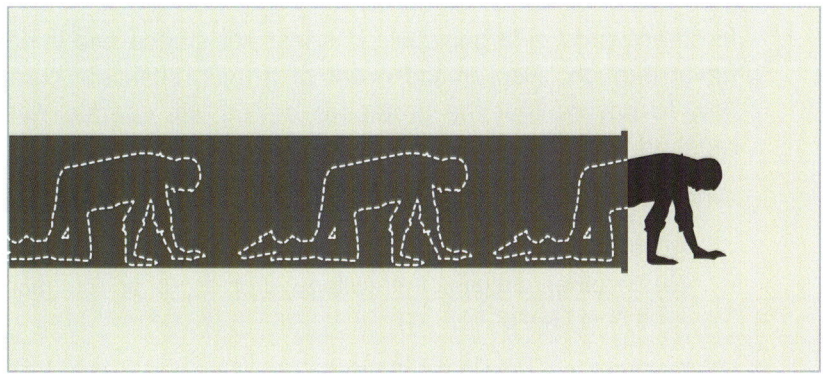

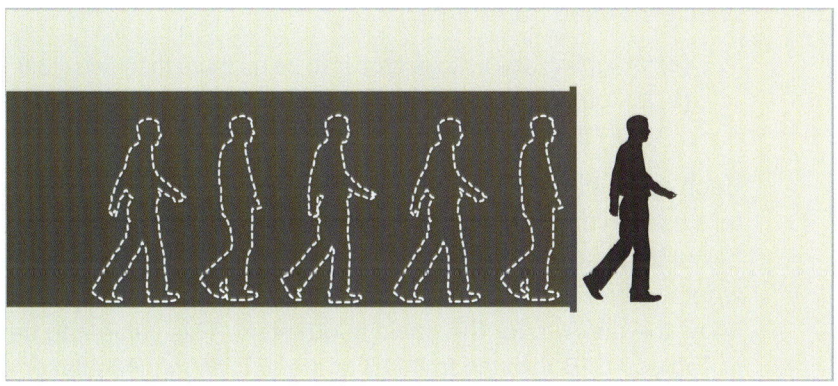

If a person lifts an object next to them weighing one-third of their body weight, they would, in most cases, not find this difficult.

If they were to crawl through a 20 m long tunnel only just big enough to fit through (high resistance), and then lift an object one-third of their weight, it would be considerably more difficult. This is because they would have already used a lot of energy to get to the end of the tunnel, leaving little energy for lifting the weight.

If they walked through a large tunnel (low resistance), then they would have more energy left to lift the weight at the other end.

A similar principle can be associated with cable. Every millimetre that is travelled uses energy. The longer the cable run, the higher the resistance and the more energy needed to get to the end.

In other words, a larger cable has less resistance and is easier to travel through. Also, in the event of a fault, sufficient current will flow to ensure that the protective device will operate within the required time.

Where there is a nominal voltage of 230 V public supply:

(a) for lighting, the voltage drop is to be no more than 3 % (6.9 V); and
(b) for other circuits, the voltage drop is to be no more than 5 % (11.5 V).

Where there is a nominal voltage of 230 V private supply (i.e. from a privately owned transformer):

(a) for lighting, the voltage drop is to be no more than 6 % (13.8 V); and
(b) for other circuits, the voltage drop is to be no more than 8 % (18.4 V).

Each cable has a millivolts per amp per metre (mV/A/m) factor. This tells us the voltage drop per amp for every metre in length, so we can calculate the overall voltage drop on a particular circuit. To establish whether the 6 mm^2 cable calculated earlier is sufficient, we need to know the mV/A/m for the cable in question. This value can be found in Tables 4D1B through to 4J4B of the IET Wiring Regulations.

For thermal insulated and sheathed flat cable with protective conductor, Table 4D5 is used.

TABLE 4D5 – 70 °C thermoplastic insulated and sheathed flat cable with protective conductor (COPPER CONDUCTORS)

CURRENT-CARRYING CAPACITY (amperes) and VOLTAGE DROP (per ampere per metre):

Ambient temperature: 30 °C
Conductor operating temperature: 70 °C

Conductor cross-sectional area	Method 100# (above a plasterboard ceiling covered by thermal insulation not exceeding 100 mm in thickness)	Method 101# (above a plasterboard ceiling covered by thermal insulation exceeding 100 mm in thickness)	Method 102# (in a stud wall with thermal insulation with cable touching the inner wall surface)	Method 103# (in a stud wall with thermal insulation with cable not touching the inner wall surface)	Reference Method A* (enclosed in conduit in an insulated wall)	Reference Method B* (enclosed in conduit on a wall or in trunking etc.)	Reference Method C* (clipped direct)	Voltage drop (per ampere per metre)
1	2	3	4	5	6	7	8	9
(mm²)	(A)	(A)	(A)	(A)	(A)	(A)	(A)	(mV/A/m)
1	13	10.5	13	8	11.5	13	16	44
1.5	16	13	16	10	14.5	16.5	20	29
2.5	21	17	21	13.5	20	23	27	18
4	27	22	27	18.5	26	30	37	11
6	34	27	35	23.5	32	38	47	7.3
10	45	36	47	32	44	52	64	4.4
16	57	46	63	42.5	57	69	85	2.8

Once **mV/A/m** has been determined, the design current (I_b) is added to the calculations.

Finally, the distance in metres (**L**) must be identified.

The equation looks like this:

$$\text{Voltage drop}(V) = \frac{mV/A/m \times I_b \times L}{1,000}$$

If the total length of cable is 10 m, the equation will look like this:

$$\text{Total voltage drop}(V) = \frac{7.3 \times 21.73 \times 10}{1,000} = 1.59 \text{ Volts}$$

As this is within the 5 % allowance for voltage drop, a 6 mm² flat twin and earth sheathed cable can be used for this installation.

Remember...

To recap, here are the steps for following the correct cable calculation procedure.

Step 1: Identify the voltage and power requirements.

Step 2: Use the calculation $I_b = \frac{P}{V}$ to identify the design current.

Step 3: Now you can identify the required rating of the protective device (I_n) using tables from Section 7 of the *On-Site Guide*. Remember that it must be equal to or higher than the design current.

7

> **Remember...**
>
> Step 4: Determine the size of the cable to be used. You can find current-carrying capacities of various cables in Appendix 4 of the IET Wiring Regulations.
>
> Step 5: Establish the voltage drop.

To calculate the voltage drop of a particular cable, you will need to look through Appendix 4 of the IET Wiring Regulations. Below the tables that show current-carrying capacity, there are tables that contain the **mV/A/m** for each cable in particular installation conditions. For example, if Table 4D1A shows the current-carrying capacity for single-core 70 °C thermoplastic insulated cables non-armoured, with or without sheathing, then Table 4D1B will show the **mV/A/m** for the same type of cable and installation condition.

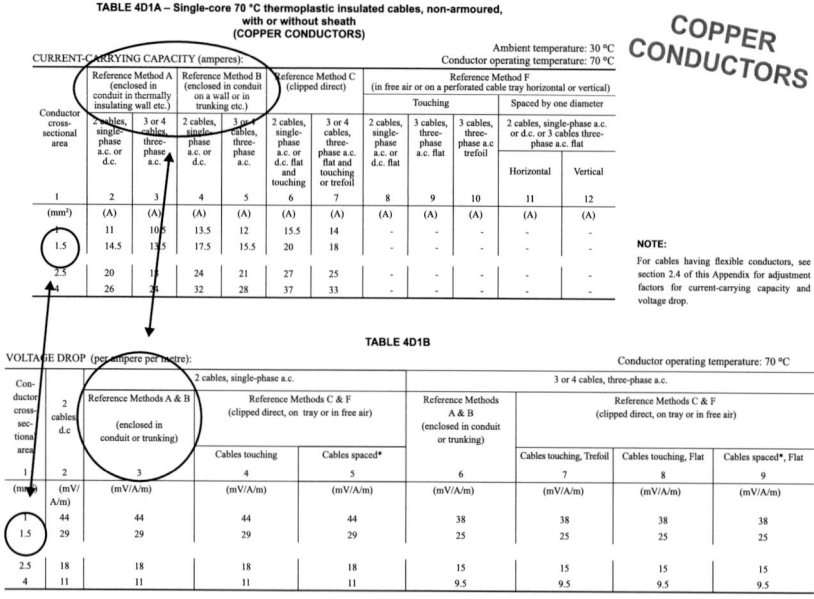

When followed step by step, the calculations are simple. The trick is not to complicate it. Complete one step before moving on to the next.

	Test your knowledge
	The following examples are intended to help you practise calculating the required size of the live conductors only.
1.	A customer would like to have an electric shower installed in their upstairs bathroom. Take a look at the following information and work out the appropriate size of 70 °C thermoplastic sheathed flat twin and earth cable and protective device required. The cable will have to run alongside the thermal insulation touching the inner wall. The protective device to be used should comply with BS EN 60898. The ambient temperature of the room is expected to be 35 °C and the total length of the cable run is 6.3 m. **(Use tables in Appendix F of the *On-Site Guide* or tables from Appendix 4 of the IET Wiring Regulations; pay particular attention to the type of cable that is to be used.)** 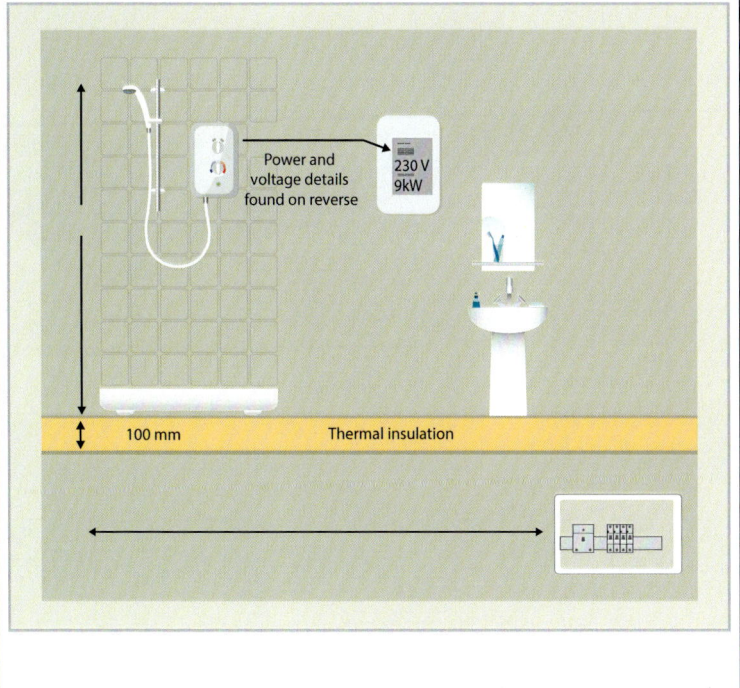

	Test your knowledge
2.	A customer requests a power supply to their shed for a light. They would like to have a 500 W security light on the outside. Using the information below, calculate the size of the cable required to supply the halogen security light. The cable to be used should be three-core armoured 70 °C thermoplastic insulated, buried direct in the ground. The ambient temperature is not expected to exceed 20 °C and the depth of burial is 0.5 m, with a soil resistivity of 2.5 K.m/W. 

Final Circuits 8

This Section provides information on the following topics:
- ▶ Designing and installing a new circuit
- ▶ Consumer unit arrangements
- ▶ Luminaires and lighting installations radial final circuits
- ▶ Ring final circuits
- ▶ Spurs
- ▶ Shower circuits
- ▶ SELV
- ▶ PELV
- ▶ FELV

8.1 General installation practices

There are certain requirements that must be complied with when making any additions or alterations to an existing electrical installation. However, just because an existing installation doesn't meet all the current requirements as set out by the IET Wiring Regulations, it does not always mean that the installation needs upgrading. For example, if an older installation has red and black insulation around the conductors instead of brown and blue, it does not mean that you cannot carry out any work on the installation without replacing all the cable in the system first. This would be impractical and would most likely upset the customer having to pay the bill.

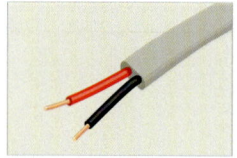

Just because it's old doesn't mean it's unsafe. You need only upgrade what is necessary to maintain the safety requirements of an electrical installation.

As an electrician, you will need to be satisfied that all the fundamental safety aspects of the installation are sufficient: for example, that protective devices are in good working order and that disconnection times can be achieved. Using the guidance available, it is the responsibility of the skilled person carrying out the inspection or installation work on an electrical system to determine if the installation can be deemed fit for continued service.

The following text can be found in the Introduction to the IET Wiring Regulations:

The Regulations apply to the design, erection and verification of electrical installations, also additions and alterations to existing installations. Existing installations that have been installed in accordance with earlier editions of the Regulations may not comply with this edition in every respect. This does not necessarily mean that they are unsafe for continued use or require upgrading.

8.2 Designing and installing a new circuit

Designing a circuit can be a long, difficult task that involves a great number of considerations, consisting of many calculations. Thankfully, the IET provides valuable information in the *On-Site Guide* and in the suite of Guidance Notes, in which many of the calculations have already been worked out for you. This will not only save you a lot of time, but will provide you with the necessary guidance so that the electrical systems you install will meet the requirements of the IET Wiring Regulations.

> ### Example
>
> Maximum EFLI values need to be obtained to ensure that a protective device operates within the required disconnection times. Below is an excerpt from Appendix B of the *On-Site Guide*, showing the predetermined EFLI values:
>
> | Circuit-breaker type | Circuit-breaker rating (amperes) | | | | | | | | | | | |
> |---|---|---|---|---|---|---|---|---|---|---|---|---|
> | | 3 | 5 | 6 | 10 | 15 | 16 | 20 | 25 | 30 | 32 | 40 | 45 | 50 | 63 | 100 |
> | 1 | 14.56 | 8.74 | 7.28 | 4.4 | 2.93 | 2.76 | 2.2 | 1.76 | 1.47 | 1.38 | 1.1 | 0.98 | 0.88 | 0.7 | 0.44 |
> | 2 | 8.4 | 5.0 | 4.2 | 2.5 | 1.67 | 1.58 | 1.25 | 1.0 | 0.83 | 0.79 | 0.63 | 0.56 | 0.5 | 0.4 | 0.25 |
> | B | 11.65 | 7.0 | 5.87 | 3.5 | 2.3 | 2.2 | 1.75 | 1.4 | 1.17 | 1.1 | 0.88 | 0.78 | 0.7 | 0.56 | 0.35 |
> | 3&C | 5.82 | 3.49 | 2.91 | 1.75 | 1.16 | 1.09 | 0.87 | 0.7 | 0.58 | 0.55 | 0.44 | 0.38 | 0.35 | 0.27 | 0.17 |
>
> For a more detailed look at earthing arrangements, see Section 6 of this Guide.
> This Section also refers to Appendix 3 of the IET Wiring Regulations, which provides information on the characteristics of various types of protective devices.

Before you get started, it is important that you assess the existing installation to determine whether or not it is capable of handling the new load or if it is suitable for alteration. If your installation is in a college or training centre workshop, the existing supply will already have been assessed for safety by the lecturer or workshop technician. However, it is always advisable to check.

8.2.1 Additions and alterations to an installation

Regulation 132.16 of the IET Wiring Regulations states that:

No addition or alteration, temporary or permanent, shall be made to an existing installation unless it has been ascertained that the rating and the condition of any existing equipment, including that of the distributor, will be adequate for the altered circumstances. Furthermore, the earthing and bonding arrangements, if necessary for the protective measure applied for the safety of the addition or alteration, shall also be adequate.

This will involve a thorough visual examination of the existing installation, as well as various tests to identify whether or not the installation is safe for continued service. An examination of the supplier's and the distributor's equipment, such as the meter, cut-out fuse, earthing arrangement, etc., will also be necessary.

One common problem that electricians may come across in their career is where an older consumer unit fitted with circuit-breakers (see BS EN 60898) or rewirable fuses (BS 3036) but no residual current devices (RCDs) is replaced with an up-to-date consumer unit. A circuit-breaker or a fuse will not detect a fault between neutral and earth and will function without any evident problem, whereas a consumer unit with RCD protection will identify this type of fault and will not allow the circuit to energize. This might mean that a neutral-to-earth fault on a circuit within the installation will not have been picked up by the old consumer unit. If the electrician has not already performed an insulation resistance test on all circuits, they may miss the neutral-to-earth fault once the consumer unit is replaced with the up-to-date version. This can lead to many complications that could have been avoided, and may increase the cost of the work being carried out, as the fault will need to be identified, located and rectified before the installation can be energized – and nobody likes unexpected cost.

8.2.2 Consumer unit arrangements

The IET Wiring Regulations require that all new consumer units installed in domestic (household) premises shall have their enclosures fabricated from a non-combustible material or be enclosed in a cabinet or enclosure fabricated from a non-combustible material complying with Regulation 421.1.201. How a consumer unit is arranged can have a significant impact on the safety of the user. For example, since the introduction of RCDs as a requirement for some electrical installations, people have been better protected from electric shock, and properties and equipment from damage arising from faults. One problem, however, is that when a single RCD has been used to protect an entire installation and a fault occurs, the RCD disconnects the supply from the entire installation. When this happens at night, for example, it creates a new hazard: it may leave the consumer with no means of illuminating their property. For those who may not be familiar with the location of their consumer unit, being left in darkness could present a serious hazard. It is therefore recommended that a consumer unit be installed in such a way that, for example, the upstairs lights and downstairs sockets will be protected by one RCD and the downstairs lights and upstairs sockets by another. This means that if one of the RCDs trips, the customer will still have power to either the lighting circuit or to the final socket-outlet circuit. This will make it very unlikely that the entire building will be left without power, and a means of illuminating either of the two floors, whether via the ceiling/wall lights or via a table lamp, would still be available.

The IET Wiring Regulations also require that the individual components within the consumer unit assembly are verified as being compatible with each other (Regulation 536.4.203). This requirement was introduced to address known safety issues relating to devices failing, sometime catastrophically, when installed in the same assembly together with other non-compatible devices. Installers need to be aware, when selecting devices for consumer units and distribution boards, that their compatibility has been verified. If an installer uses various devices not specified by the manufacturer(s) as being compatible, then they (the installer) will become the assembly manufacturer and adopt the associated responsibilities.

▼ **Figure 8.1** Recommended arrangement for an RCD protecting a domestic installation

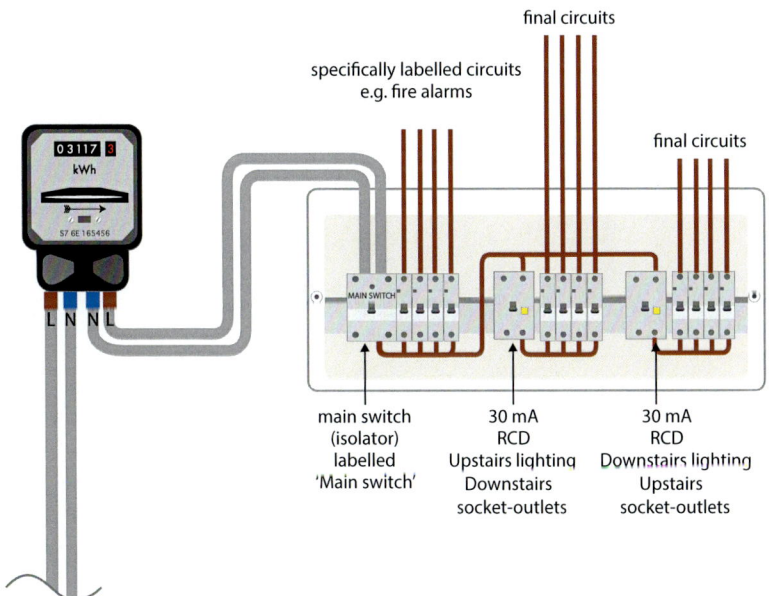

8.2.3 Heights of switches, socket-outlets and consumer units

When planning any new installations, switches and socket-outlets must be easily accessible to all persons and should therefore be installed within the required heights. To comply with Building Regulations, light switches and socket-outlets must be installed between 450 mm and 1,200 mm from the finished floor level. Consumer units are required to be installed between 1,350 mm and 1,450 mm above the finished floor level. If you are working on an older installation that does not meet these requirements, it is **not** necessary to re-position all switches, socket-outlets and the consumer unit to these heights.

NOTE: The requirements applicable to dwellings for known wheelchair users will vary.

▼ **Figure 8.2** Required heights for switches, socket-outlets, etc.

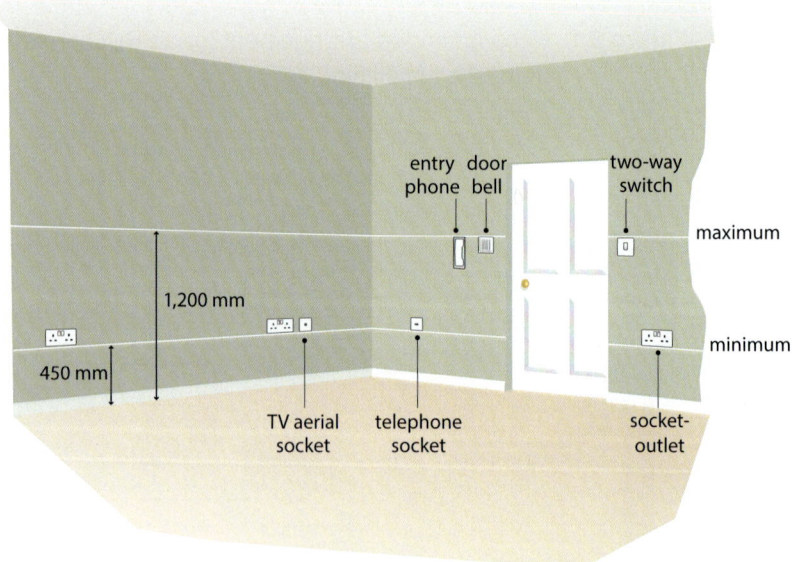

With any new installation, it should be determined if additional protection is needed by means of an RCD (see Regulation 411.3.3).

When installing cables, Chapter 52 and, in particular, Section 522 *(Selection and erection of wiring systems)* should be considered. These regulations cover the requirements for protecting cables against damage when they are installed within walls and partitions. There may be factors to take into consideration such as metal parts that may become live if the cable is damaged and live conductors become exposed.

8.2.4 Cable routes

You know where to mount your accessories – now you need to consider how you will get the cable to them. You could save time and money by running cables in straight lines to where you need them. But you need to consider what someone might think when putting up a picture frame or similar. In most cases, they would not install this above or below a switch or socket-outlet: it is expected that they would consider the possibility that there might be a cable buried in the wall directly above, below or to either side of the equipment. In an ideal world, therefore, they would choose a position outside of these areas to hang the family portrait; however, this cannot be guaranteed!

Figure 8.3 shows the zones where cable routes are permitted, providing they have a minimum degree of protection. Full details of the minimum degrees of protection required can be found in Section 7 of the *On-Site Guide*.

▼ **Figure 8.3** Permitted cable routes

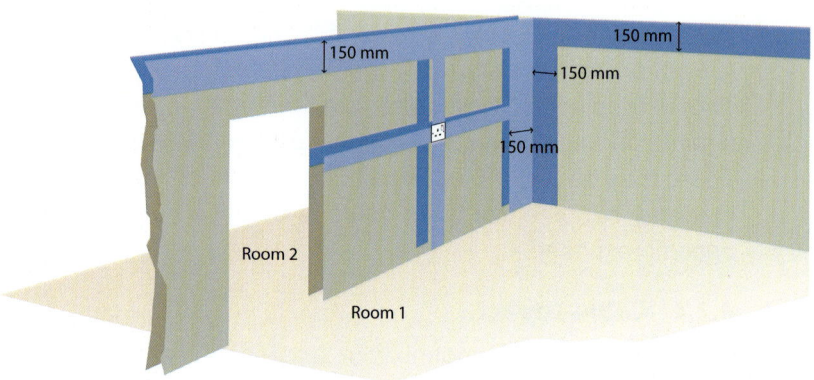

The aspects we have discussed are some of the fundamental ones that should be considered prior to installing any electrical installation. There are many, many more that you will become familiar with as you gain more experience, studying and working in the electrotechnical industry. These include, for example, the question of how far away from the top and bottom of joists holes need to be, when drilling joists to run cables through, and the maximum depth of chases, when chasing walls. If these things are not done correctly, there may be serious consequences, such as the house falling down! Keeping this in mind and always asking a supervisor will help you avoid making mistakes.

The IET Wiring Regulations require the provision of adequate support for wiring systems so as to prevent their premature collapse in the event of a fire. This requirement covers all wiring systems, whether or not they are in escape routes (Regulation 521.10.202) and was introduced as a result of known safety issues that have resulted in the deaths of several firefighters.

BS 7671:2018+A2:2022 introduced a definition for 'protected escape routes' and the requirements for these specific locations. Regulation 422.2 requires that cables or other electrical equipment shall not be installed in a protected escape route unless they are part of an essential fire safety or related safety system, general needs lighting or socket-outlets provided for cleaning or maintenance.

This means that, in general, cables in a protected escape route should be limited to lighting, emergency lighting and fire detection and alarm systems, although cables for other safety systems may be necessary. Hospitals may have special requirements as detailed in Section 710 of the IET Wiring Regulations.

Regulation 422.2.1 identifies the requirements for the types of cables and cable management systems which may be installed in protected escape routes.

Further guidance on protected escape routes is provided in Appendix 13 of the IET Wiring Regulations.

8.3 Luminaires and lighting installations

8.3.1 Regulations and guidance

Section 559 of the IET Wiring Regulations provides the requirements for designing and installing a lighting circuit. Before you can start work on such an installation, you must find out what the customer would like. There are a number of questions that will need to be answered, such as:

(a) How many luminaires are to be installed?

(b) What kind of lamps will be used in the light fittings? (For example, LED (light-emitting diode), fluorescent, halogen, incandescent, etc.)

(c) Are they to be installed indoors?

(d) Are they to be installed outdoors? Section 714 of the Wiring Regulations provides the requirements for outdoor lighting installations.

(e) What are the manufacturer's requirements for the luminaires?

(f) Will they all be switched on and used at the same time for long periods of time, as in an office building or a factory? (See Section 12 of this Guide on diversity.)

(g) Will they be in a dwelling and not in continuous use? (See Section 12 of this Guide on diversity.)

(h) What type of installation method will be used? (See Table 4A2 of Appendix 4 of the IET Wiring Regulations.)

(i) What type of switches are required, i.e. one-way, two-way or two-way and intermediate?

(j) Does the installation comply with the relevant parts of the Building Regulations?

These are important factors to keep in mind, as they can have a significant effect on the design and cost of the installation.

At this point, if any of the customer's requests do not comply with the IET Wiring Regulations, the electrician must advise accordingly. For example, if a customer supplies a light fitting that does not comply with the relevant product standard, such as a fitting purchased from overseas with no manufacturer's documentation, and requests that it be installed, the installer must advise the customer that the fitting does not comply with Regulation 559.3.1 and the light fitting should not be installed.

8.3.2 Selection of equipment

Protective device

The first part of a lighting system is the consumer unit, which has been described earlier in Section 8.2.2.

The consumer unit will have an appropriate protective device for the lighting circuit. The rating of this protective device will have been determined when calculating cable sizes. When lighting circuits are installed in a domestic (household) premises, they will be provided with additional protection by means of an RCD with a residual operating current not exceeding 30 mA, as required by Regulation 411.3.4.

▼ **Figure 8.4** Protective device

Luminaires

The type and characteristics of a luminaire will vary, depending upon the environment in which it is to be installed. For example, if the lighting is required in a swimming pool, it must have the required IP rating to be submerged underwater. If it is being installed in a workshop, the ability for the fitting to withstand mechanical impact should be considered. The IP rating refers to the requirements of the IP code, which uses a two-part numerical system to define the level of protection provided by an enclosure. The first numeral is related to the protection against solid bodies and the second numeral to the protection against liquids. Details of the IP code and the requirements are contained in Appendix C of this Guide.

Ceiling rose

▼ **Figure 8.5** Terminals at the ceiling rose

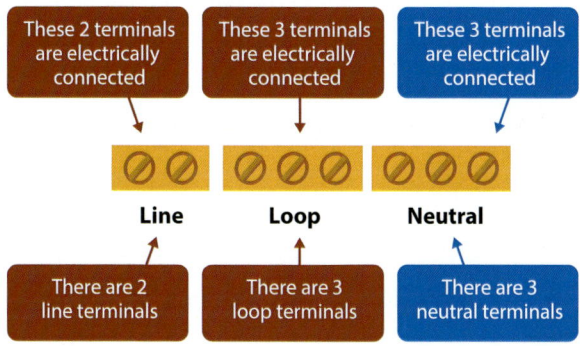

NOTE: Circuit protective conductor (cpc) terminals not shown for clarity.

8

Power to a lighting circuit is usually supplied by two different methods. These are commonly referred to as the two-plate and the three-plate methods.

The two-plate method

This method requires the supply to go straight to the switch, where, in an installation that uses two-core and earth, only the line conductor is switched and the neutral continues through the switch to the luminaire. It is common for the neutral conductors to be connected together at the switch using a connector that complies with the requirements of the IET Wiring Regulations. By introducing a new connection, you are essentially creating a new point that could introduce a safety risk. This must be borne in mind when selecting suitable components to terminate the conductors; you must also check that they comply with the relevant standards. The two-plate method is growing in popularity and is the preferred method when using certain types of switches that need to have a neutral in order to function correctly, such as smart switches that can be controlled remotely using 'smart home' apps.

▼ **Figure 8.6** Two-plate method

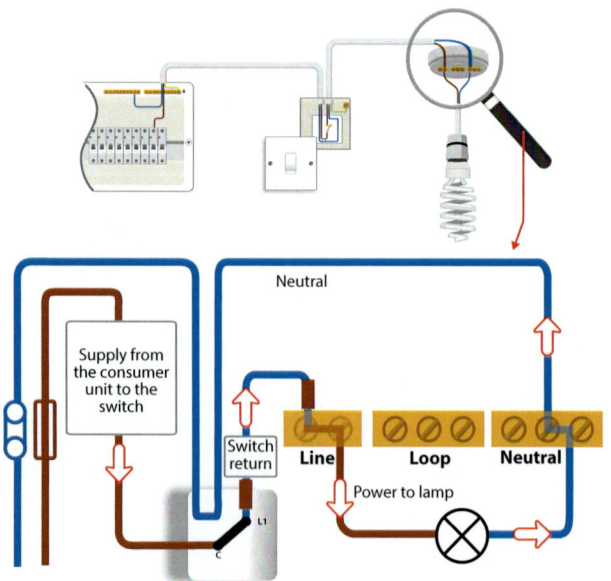

The three-plate method

This method of wiring involves the supply to the ceiling rose, a switch wire to the switch and a return switch wire back to the luminaire. All three connection points of a three-plate ceiling rose are used, for example, line, loop and neutral.

▼ **Figure 8.7** Three-plate method

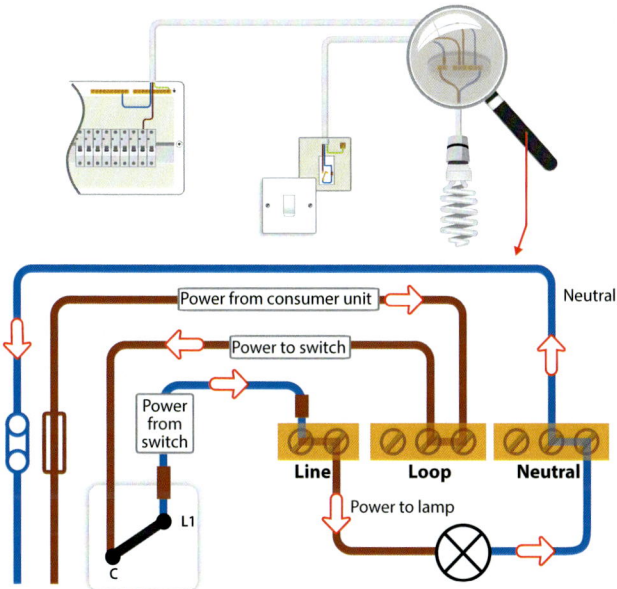

Use of the three-plate method is common when more than one light needs to be operated by separate switches on a circuit. It enables the installer to provide power to multiple lighting points that can be switched individually. In the past, this method was commonly used in domestic premises for ease of installation, as it allows you to run a single cable to supply all light fittings and then just one cable to the switch from each point. The primary supply for the circuit comes into the ceiling rose and can be looped to another ceiling rose, and to as many more as needed, provided that the current-carrying capacity of the cable is not exceeded, the permitted voltage drop is not exceeded and disconnection times can be achieved. This is shown in Figure 8.8.

▼ **Figure 8.8** Looping power

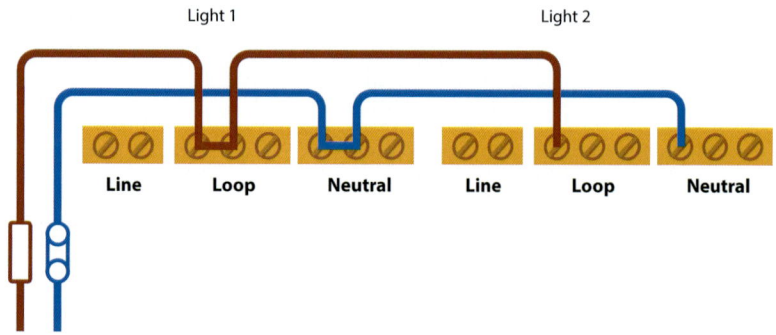

Cable

The line conductor can be either a single-core cable or contained within a multi-core cable. The size and type of the cable will be determined using a cable calculation that takes into account the overall load of the luminaires required, as well as the environment into which it is to be installed.

The cpc is not shown in the following diagrams, as these simply represent the functionality of the circuit. However, when installing any electrical system, it should be determined what is needed in the way of a protective conductor (see Section 6).

Switching devices

Depending on the requirements of the system, there are various switching devices that can be used. It is important to consider the type of fittings that are going to be installed and to refer to manufacturers' instructions, so that the correct switches can be selected for the installation. Some lamps and their controlgear draw a lot of current when they are first switched on and this needs to be considered when selecting the switching device. The amount of current that switches are designed to use can vary. If an underrated switch that cannot handle the current is installed, there is a risk of the switch being damaged and potentially overheating, leading to a fire.

One-way switch

If the light needs to be switched on and off from only one place, then a one-way switch, like the one below, will be required.

▼ **Figure 8.9** One-way light switch

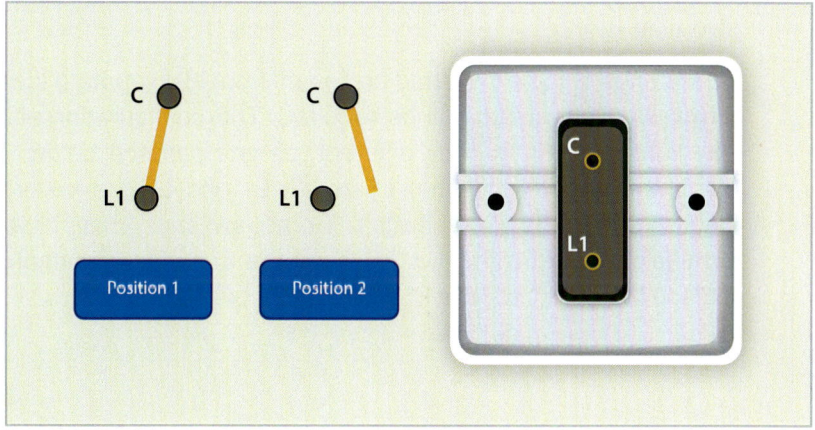

The following image shows a simple one-way switch wiring configuration. This is a typical method used that requires a light to be controlled from only one position.

▼ Figure 8.10 Connection of conductors in a one-way light switch

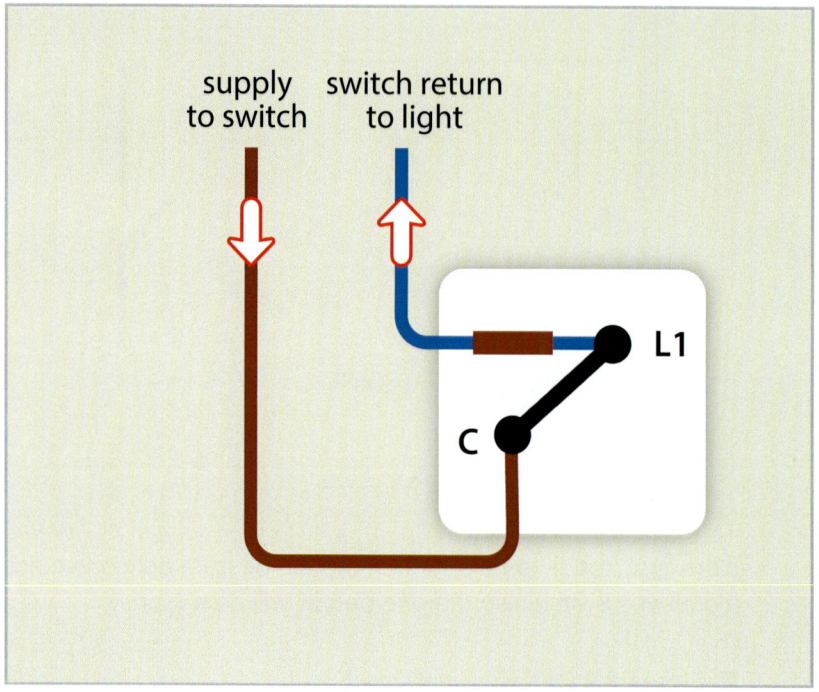

Two-way switch

If a light needs to be switched on and off from two separate locations, a two-way light switch will be required. The configuration of a two-way light switch is slightly different, as you can see below. The C (common) will link to either L1 or L2. L1 and L2 are never linked together on a two-way switch. Some switches may have their terminals marked with L1, L2, L3, as the manufacturer chooses to replace the C with L1, the L1 with L2 and the L2 with L3.

▼ **Figure 8.11** Two-way light switch

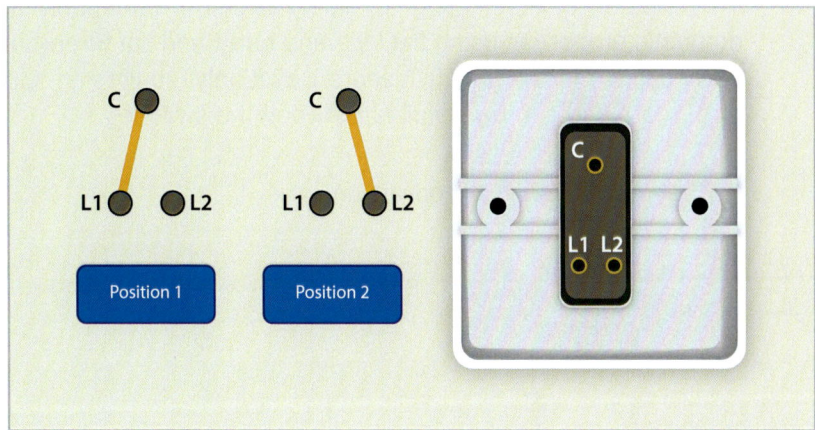

The following image shows a two-way conventional wiring configuration. This method of connecting a lighting circuit is commonly used in commercial or industrial environments using single-core cable.

▼ **Figure 8.12** Two-way conventional

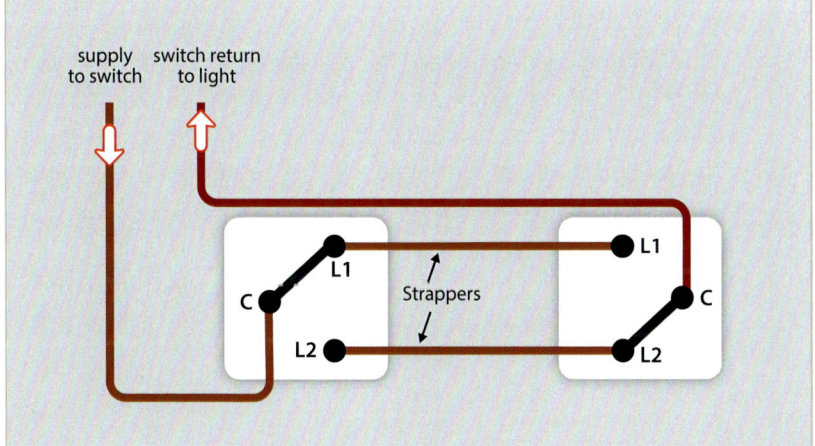

8

The following image shows a two-way conversion wiring configuration. This method of connecting a lighting circuit is commonly used in domestic properties using flat twin and earth and flat three-core and earth cable. This method is favoured, as it only requires an additional three-core cable to be installed between the switches.

▼ **Figure 8.13** Two-way conversion

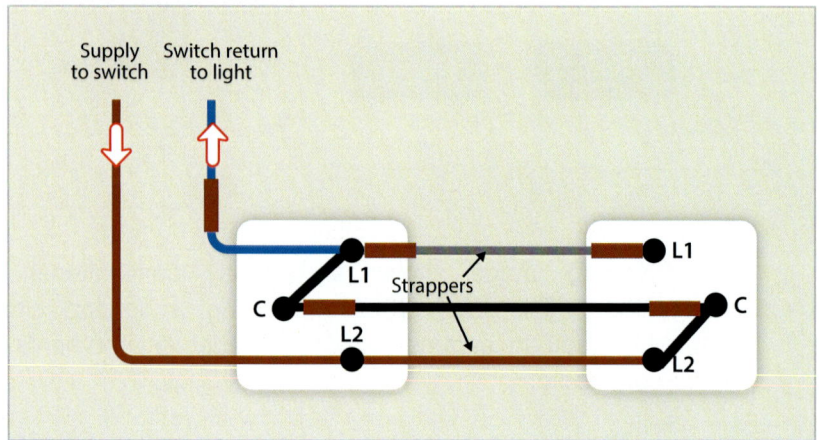

Intermediate switch

If the lighting system needs to be switched on and off from three or more locations, such as in a stairwell rising over several floors, an intermediate switch will be required. Looking at the figure below, you can see how the terminals are connected to one another during operation of the switch.

▼ **Figure 8.14** Intermediate light switch

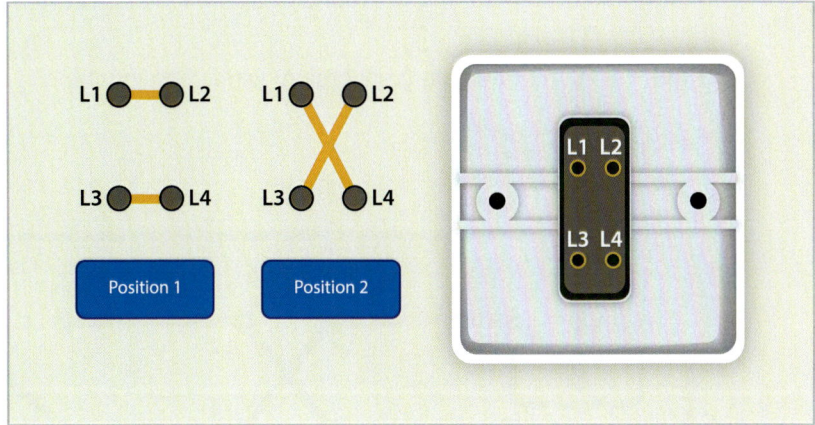

The following image shows an intermediate conventional wiring configuration commonly used in commercial and industrial installations.

▼ **Figure 8.15** Intermediate (conventional) wiring configuration

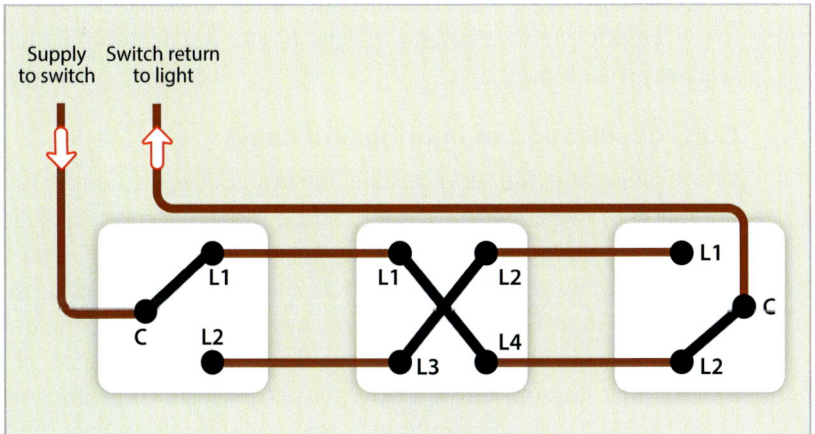

The following image shows an intermediate conversion wiring configuration that is commonly found in domestic installations.

▼ **Figure 8.16** Intermediate (conversion) wiring configuration

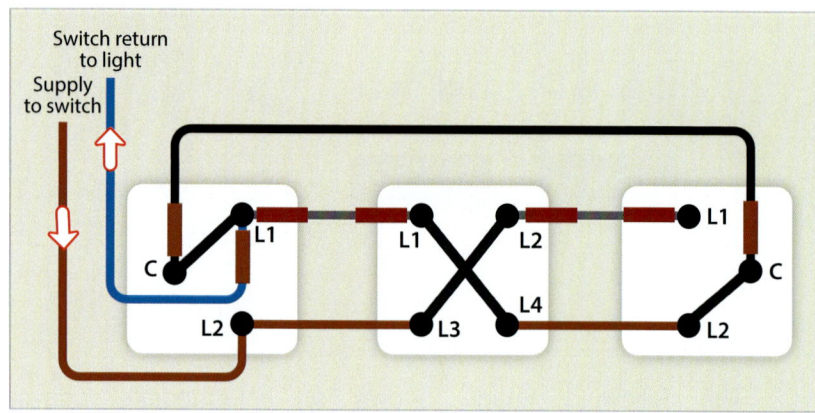

8.3.3 Installation conditions

There are not many places in the developed world where you won't find lighting of some sort – from ambient lighting for effect (such as pond lighting) to functional lighting for illuminating large areas for a workforce to operate safely (such as portable lighting for highway maintenance).

Correct selection of lighting and cable

When selecting the appropriate lighting for a particular task, the condition of the environment should be taken into consideration, not only for the luminaire and the light it can provide, but also for the entire wiring system. In Section 1 of this Guide, there is a list of the cables that can be used for wiring systems. Some are suitable for small domestic installations where there will be little risk of heavy mechanical impact. Others have varying levels of protection, from steel braided wire to complete metal sheathing, such as steel-wire armour (SWA) or mineral-insulated metal-sheathed (MIMS) cable, that will withstand significant mechanical impact. Of course, these vary dramatically in price and the time it takes to install. Usually, more skill and time is required for the installation of cable that has increased resistance to mechanical damage.

The light fitting to be installed should also be chosen to suit the environment. To best select a light fitting, an assessment of the environmental conditions should be made – not only of these conditions during installation, but also of what the area will be used for after the installation is complete.

For example, if you are installing a lighting circuit in a sauna, you will require a wiring system that can withstand the moisture and high temperatures expected in that environment.

Access

Lighting is generally installed at high level, so some form of access equipment, such as a stepladder, will be needed. More complicated access equipment, such as a tower or scaffolding, can add considerable cost and time to the installation and customers should be informed of this prior to work commencing.

> **Remember...**
> When connecting conductors into terminals, use a wiring or circuit diagram and focus on one conductor at a time. When working with two-way switches, it is good practice to connect your strappers first: L1 at one switch should be connected to L1 at the other switch and L2 at one switch should be connected to L2 at the other switch (when flat twin and earth cable is used for switch wires, the line conductor needs to be identified with brown sleeving). Remember that there should never be a neutral conductor connected to a standard one-way, two-way or intermediate switch for lighting. As mentioned earlier, there are switches that require a neutral to function, but these are generally for lighting control gear or smart device installations.

8.4 Radial final circuits (power)

8.4.1 Regulations and guidance

Radial final circuits for socket-outlets are not as common in the UK as ring final circuits. However, some electricians prefer to use radial final circuits to supply socket-outlets. Appendix 15 of the IET Wiring Regulations provides guidance and information on the installation of radial final circuits used for socket-outlets.

A radial circuit is a simple circuit to install. A 230 V radial circuit will consist of two live conductors (line and neutral) and a cpc. These will be fed from an appropriate protective device, such as a circuit-breaker or fuse and RCD.

8.4.2 Selection of equipment

The circuit will start at the consumer unit and this is where a suitable protective device will be installed. In new installations, this is likely to be a circuit-breaker (see BS EN 60898) or a residual current circuit-breaker with integral overcurrent protection (RCBO) (see BS EN 61009). The cable used will be sufficient to handle the expected current that will be flowing through the circuit: guidance from Table H2.1 of the *On-Site Guide* can be used. However, it should be borne in mind that installation conditions may affect the current-carrying capacity of the cable. The socket-outlets should be manufactured to BS 1363. Regulation 553.1.3 and Table 55.1 from the IET Wiring Regulations provide more information on the minimum requirements for plugs and socket-outlets for low voltage (LV) circuits. There are socket-outlets available that have USB (universal serial bus) circuits incorporated into them. BS 1363 states that USB circuits incorporated in a socket-outlet shall conform to the relevant standards for information and technology equipment. There are also socket-outlets manufactured to BS 7288 which incorporate an RCD.

8.4.3 Maximum number of socket-outlets

Although there is no specified maximum number of socket-outlets that can be installed on a final radial circuit, the circuit should meet the requirements of Regulation 433.1, and further guidance is given in Appendix 15 of BS 7671. This information is summarized in Section 8.4.5 of this Guide.

Remember: if a cable passes through or runs alongside thermal insulation, it significantly reduces the current-carrying capacity of the cable (see Section 7 Cable Calculations).

8.4.4 Minimum and maximum heights for socket-outlets and switches

To comply with the Building Regulations, socket-outlets in new installations must be installed above 450 mm and below 1,200 mm. This is to ensure equality of access, allowing people with limited reach, such as wheelchair-users, to access switches and socket-outlets. Regulation 553.1.6 of the IET Wiring Regulations requires socket-outlets to be positioned in locations to allow plenty of clearance from worktops, shelves and other objects or furnishings, so that damage to the flexible cable and the plug of any equipment when being plugged in or unplugged is prevented.

8.4.5 Maximum area that a radial final circuit can serve

A 30/32 A radial final circuit using 4 mm^2 copper conductor thermoplastic or thermosetting insulated cables should serve an area not exceeding 75 m^2. A 20 A radial final circuit using 2.5 mm^2 copper conductor thermoplastic or thermosetting insulated cables should serve an area not exceeding 50 m^2.

Table H2.1 in the *On-Site Guide* provides information on the required csa of conductors and the maximum area that a radial final circuit can serve, as well as the rating of the protective device.

8.5 Ring final circuits

The definition of a ring final circuit (from Part 2 of the IET Wiring Regulations) is:

> *A final circuit arranged in the form of a ring and connected to a single point of supply.*

8.5.1 Regulations and guidance

Ring final circuits are used to supply accessories such as socket-outlets complying with BS 1363-2 and fused connection units complying with BS 1363-4, and are almost unique to the UK. In 1942 the Electrical Installations Committee determined that the ring final circuit offered a more efficient and cheaper wiring system, which would support a greater number of socket-outlets. Section 553.1 of the IET Wiring Regulations covers the requirements for socket-outlets installed on a ring final circuit.

Regulation 433.1.204 provides details of the type and rating of the protective device that must be selected, the minimum csa of the line conductors, and the cpc, including the minimum cpc. Appendix 15 of the IET Wiring Regulations and Appendix H of the *On-Site Guide* provide guidance on ring final circuits.

The ring final circuit is intended to evenly distribute the load on the circuit, so that the maximum current-carrying capacity of the cable is not exceeded for long periods of time.

The illustration below shows how the load is distributed evenly over two legs of the ring (the current flow is shown with the red arrows). This reduces the strain on the cable and the risk of it being overloaded.

▼ **Figure 8.17** Even distribution of load over a ring final circuit

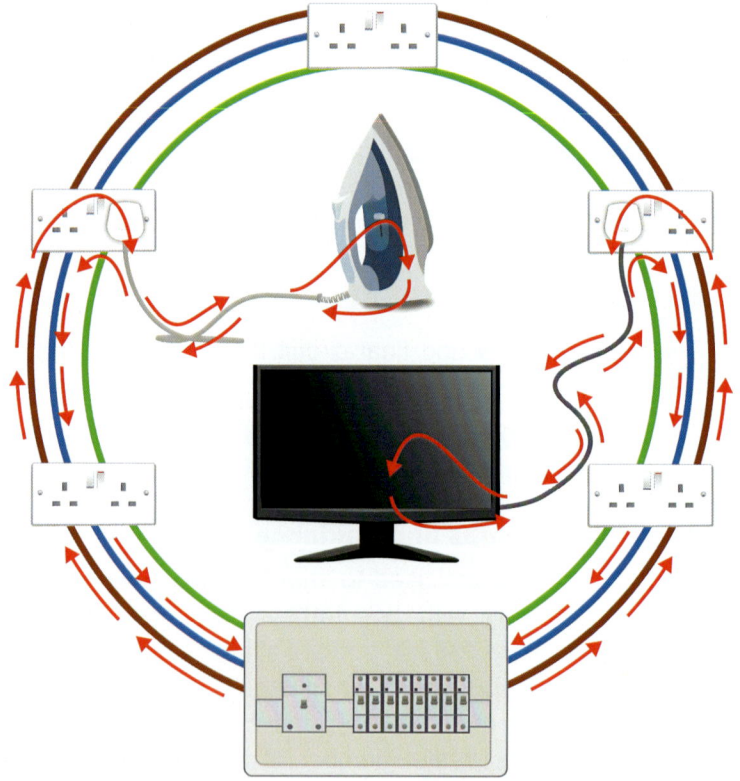

8.5.2 The required csa for conductors on a ring final circuit

The csa of the live conductors (line and neutral) for a ring final circuit should be 2.5 mm^2 and the cpc should be no less than 1.5 mm^2.

8.5.3 Maximum number of socket-outlets

An adequate number of socket-outlets should be provided so that equipment can be supplied from a conveniently accessible socket-outlet, taking into account the length of flexible cable normally fitted to portable appliances (see Regulation 553.1.7). This basically means that you can have (within reason) as many socket-outlets as required for the installation. The IET Wiring Regulations do not state a maximum number of socket-outlets that can be installed on a ring final circuit. However, Regulation 433.1.204 clearly states that, given the intended condition of use, it should be unlikely to exceed for long periods of time the current-carrying capacity (I_z) of the cable. Consequently, the responsibility sits with the designer to determine the number of socket-outlets to be installed.

8.5.4 Maximum area that a ring final circuit can serve

A ring final circuit should serve an area not exceeding 100 m^2. Table H2.1 in the *On-Site Guide* provides information on the required csa of conductors and the maximum area that a ring final circuit can serve, as well as the rating of the protective device and the minimum csa of the cpc. This table can also be referred to when looking at radial circuits.

▼ **Figure 8.18** Taken from Table H2.1 in the *On-Site Guide*

			Minimum live conductor cross-sectional area* (mm^2)		
Type of Circuit		Overcurrent protective device rating (A)	Copper conductor thermoplastic or thermosetting insulated cables	Copper conductor mineral insulated cables	Maximum floor area served (m^2)
1	2	3	4	5	6
A1	Ring	30 or 32	2.5	1.5	100

8.6 Spurs

8.6.1 What is a spur?

A spur is simply an individual cable coming from a point on a ring or radial circuit to supply an accessory or piece of equipment. This can be a socket-outlet or a supply to equipment. There are two types of spurs: unfused and fused. Both are added to a final circuit by connecting a cable into an existing socket-outlet or joint box and then to the new socket-outlet or fused connection unit.

8.6.2 Adding a spur to a circuit

A spur is an amendment or addition to an existing circuit that is permanently connected directly to a ring final circuit or radial circuit.

When adding one single or double socket-outlet to a final socket-outlet circuit, there is no requirement to install a fused connection unit, providing that only one single or one double socket-outlet is connected to a single point of the circuit. This is to prevent any single branch becoming overloaded.

If more than one socket-outlet is required to be installed from a single point on the ring final or radial final circuit, additional protection is required by means of a fused connection unit manufactured to BS 1363-4, as shown in Figure 8.19. Providing the requirements of Regulation 433.1.204 are met, many socket-outlets or permanently connected items of equipment can be fed from the existing ring final circuit. However the total number of non-fused spurs should not exceed the number of socket-outlets or items of stationary equipment connected directly into the circuit.

▼ **Figure 8.19** Ring final circuit with spurs

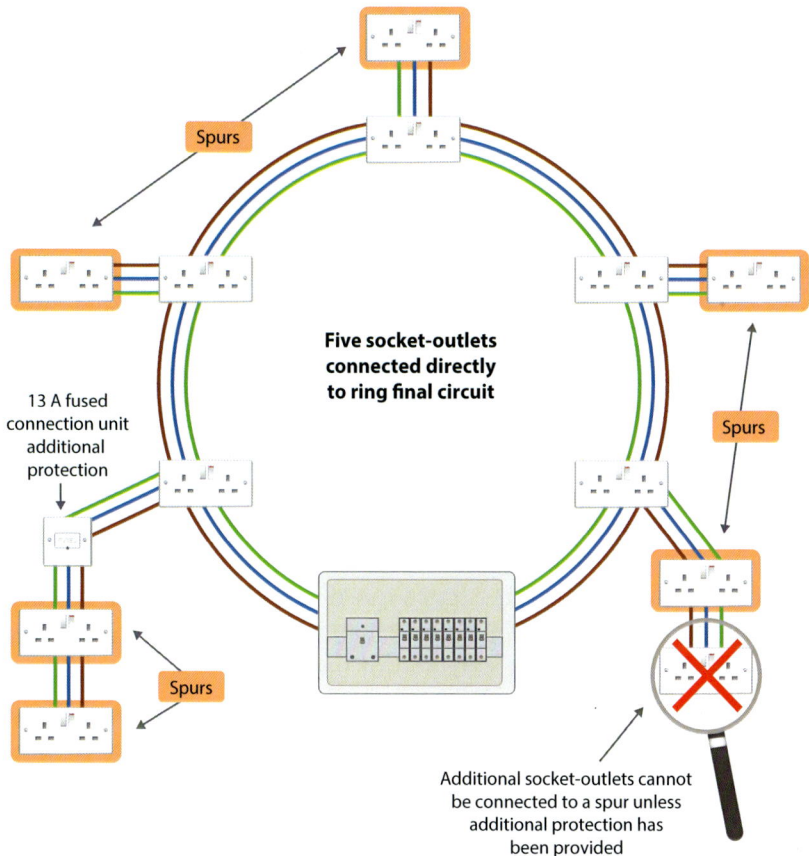

▼ **Figure 8.20** Connection of the cpc to a socket-outlet including earth tail

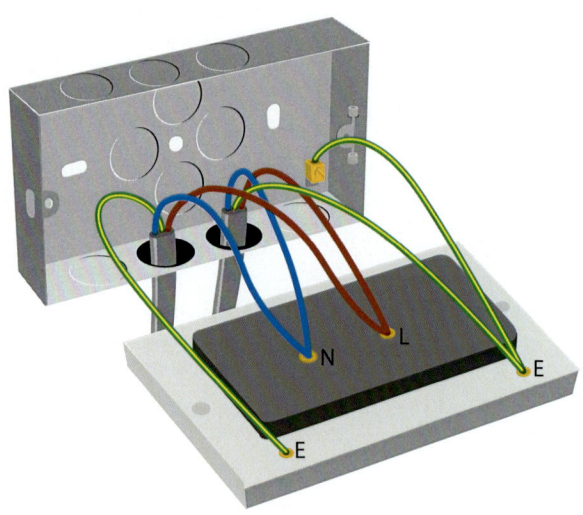

By ensuring that the cpcs have been terminated correctly, the general safety of the circuit will be improved, especially under fault conditions. The construction of the socket-outlets is such that the socket-outlet earth terminals are permanently linked together.

8.7 Shower circuit

One of the simplest circuits that you will ever have to install is one that only serves one piece of equipment, such as a shower. Although there may be processes that are more complicated going on inside the equipment, the supply to the equipment is straightforward.

8.7.1 A circuit supplying a shower

Each circuit supplying a shower will require:

  (a) appropriately selected and rated protective devices for overload, short-circuit and earth fault protection;
  (b) cable with suitable current-carrying capacity;
  (c) an isolator; and
  (d) the equipment itself, i.e. the electric shower.

A circuit for an electric shower shall be protected by an RCD in accordance with Regulation 701.411.3.3 of the IET Wiring Regulations, with the characteristics specified in Regulation 415.1.1, and be designed to supply the full load. Diversity shall not be applied on this type of circuit (see Section 12 of this Guide).

8.8 Separated extra-low voltage systems (SELV)

For the regulations related to SELV, see Section 414 of the IET Wiring Regulations.

▼ **Figure 8.21** SELV transformer

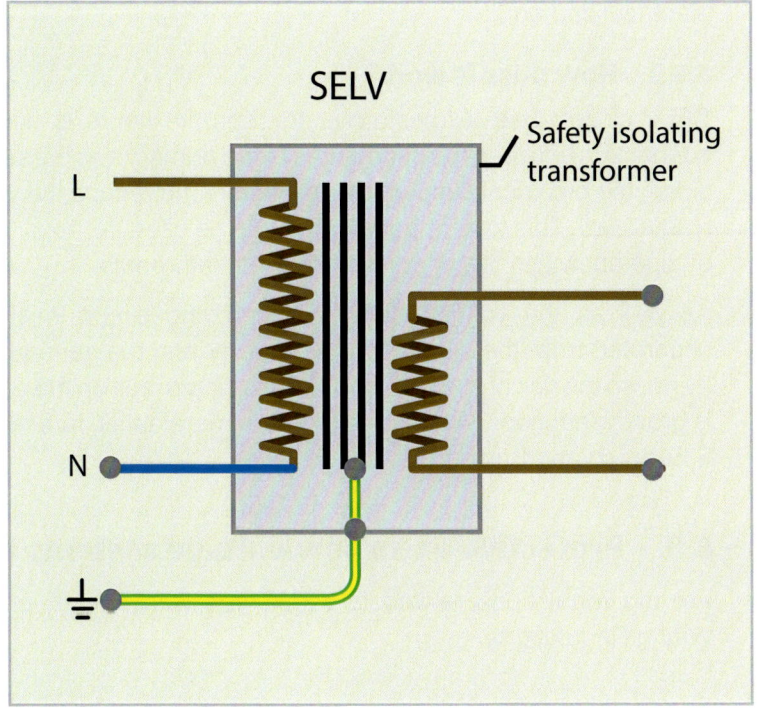

8.8.1 BS EN 61558-2-6 safety isolating transformer

For circuits that require further protection so that the risk of electric shock is reduced, for example, extractor fans installed within a location containing a bath or shower, a SELV supply can be used. This type of system separates incoming supply to the property from the supply to the equipment by using an isolating transformer. This type of transformer not only reduces the voltage to supply the equipment, but also completely separates the electrical supply from the equipment. Figure 8.21 clearly shows that there is no physical electrical connection between the primary and secondary conductors. The power that is supplied to the equipment is produced by mutual inductance (you will learn more about this process throughout your electrical course).

8.8.2 How does it work?

SELV circuits use a transformer to separate (there is no physical connection between the conductors of the primary and the secondary sides) the electrical supply from the circuit and reduce the voltage to a safe level. This type of transformer can be used to supply electrical equipment within the permitted zones in bathrooms.

As well as the voltage being reduced, the circuit is electrically separated from the supply to the property and the general mass of Earth, so the risk of electric shock from a single fault on the equipment is greatly reduced. To understand this in more detail, refer to Section 6 Earthing and Bonding.

8.9 Protective extra-low voltage systems (PELV)

For the regulations related to PELV, see Section 414 of the IET Wiring Regulations.

▼ Figure 8.22 PELV transformer

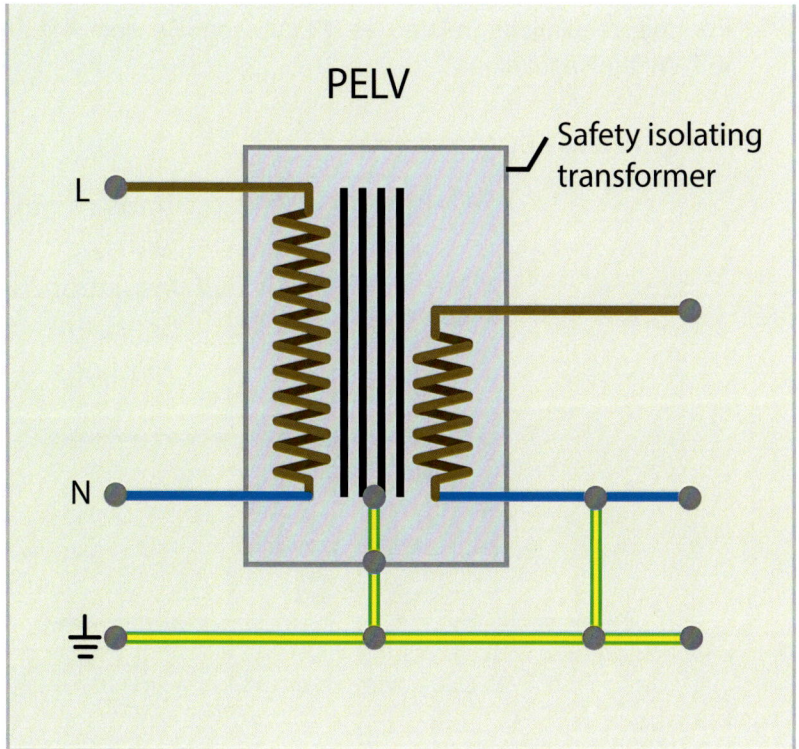

8.9.1 BS EN 61558-2-6 safety isolating transformer

This type of isolating transformer has all the characteristics of SELV, except that it does not provide electrical separation from earth. This type of circuit should be considered when the characteristics of SELV cannot be achieved and maintained throughout the life of the installation. A typical example of where you would find a PELV transformer on a circuit is in a computer.

As you can see from Figure 8.22, the secondary winding of the transformer has its neutral connected to earth.

8.10 Functional extra-low voltage systems (FELV)

For the regulations related to FELV, see Section 411.7 of the IET Wiring Regulations.

▼ **Figure 8.23** FELV transformer

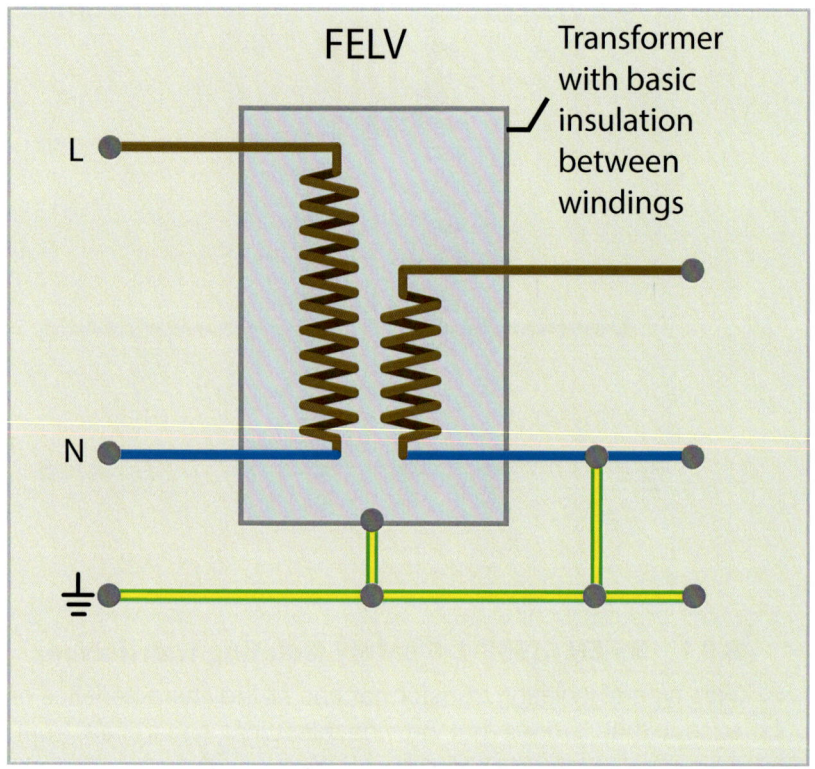

A transformer that has the characteristics of FELV uses extra-low voltage (ELV). However, it is not adequately protected from accidental contact with higher voltages in other parts of the circuit. FELV is used where ELV is required for functional purposes, such as for machine control systems.

	Test your knowledge
1.	What are the two different types of spur connections which can be made to a ring final circuit?
2.	Using Table H2.1 from the *On-Site Guide* (referred to earlier in this section), what are the two overcurrent protective device ratings that can be used for a ring final circuit?
3.	Three additional double socket-outlets are to be installed on one branch of a spur from a ring final circuit. What is required, with regard to overload protection, so that the additions meet the requirements of the IET Wiring Regulations?
	For the following questions, see Section 559 of the IET Wiring Regulations.
4.	What is the maximum permitted operating voltage of a lighting circuit that includes a ceiling rose or lampholder?
5.	Which regulation deals with stroboscopic effect?

	Test your knowledge
6.	Have a look at the following diagrams and state whether the light will be on or off. 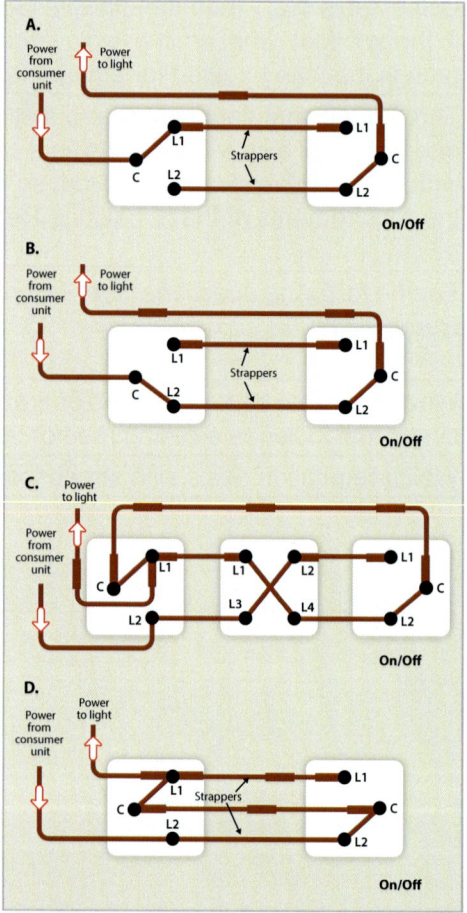
7.	Describe two characteristics of SELV circuits that protect the user from electric shock.
8.	What is the difference between a SELV transformer and a PELV transformer?
9.	What section of Part 7 of the IET Wiring Regulations should be referred to when installing a shower?
10.	Where might you find FELV transformers?

Inspection and Testing 9

This Section provides information on the following topics:
- ▶ The reason for testing
- ▶ Initial inspection
- ▶ Testing procedures

9.1 Why inspect and test?

The reason behind inspecting and testing an electrical installation is to find out whether a new installation is safe to be energized and handed over to the client, or whether an existing installation is safe to remain in service until the next inspection is due. The Inspector must verify that the existing installation, including the equipment that belongs to the distribution network operator (DNO) or the supplier, is adequate and safe. If any of the existing equipment does not support the planned new installation or is in poor condition, the installer should not proceed.

NOTE: The term 'Inspector' is used throughout this Section, and this refers to the person(s) carrying out the inspection and testing of the installation. This is usually, but not necessarily, the same person.

9.1.1 Legal aspects

The person(s) carrying out the inspection and testing must comply with the Electricity at Work Regulations 1989 (EAWR). The EAWR require that whoever does the work has to be competent to do so. When testing electrical installations, some of the tests are carried out on live parts, so it is extremely important that the Inspector is competent to carry out this work. Installations that comply with BS 7671 will, in most circumstances, meet the requirements of the EAWR. Note that electrical contractors who issue inaccurate or false Electrical Installation Certificates (EICs) may be liable to prosecution.

9.1.2 Responsibilities of the Inspector

The Inspector must:

(a) be competent to do the work, as required by Regulation 16 of the EAWR;
(b) be familiar with the requirements of the most recent version of the IET Wiring Regulations;
(c) have a good knowledge of the type of installation that they are inspecting;
(d) understand the inspection and testing procedures; and
(e) use testing equipment manufactured to the required standards and in accordance with HSE Guidance Note GS38 *Electrical test equipment for use by electricians*.

It is the responsibility of the Inspector to:

(a) identify that reasonable measures are in place to reduce the risk of injury to persons or livestock;
(b) identify that reasonable consideration has been given to installation methods so that the risk of damage to property as a result of an electrical installation is reduced;
(c) compare the inspection and testing results with the design specifications; and
(d) make a judgement on the condition of the installation and advise on remedial works to the client.

Where the inspection highlights an extremely dangerous situation, for example, where persons or livestock are at immediate risk of electric shock or an imminent fire hazard is evident, then urgent action is advised to remove the danger. As an expert, the Inspector has been employed to identify electrical problems and therefore is advised to exercise judgement to secure the area and inform the client immediately, followed up in writing. Once permission has been obtained, the danger should be removed by isolating the circuit or circuits in question.

9.1.3 Why is initial verification important?

The reason initial verification is carried out on an electrical installation is to confirm, by way of visual and physical inspection, as well as testing, during construction and on completion, that the installation complies with the design and the requirements of the IET Wiring Regulations, as well as the relevant parts of the Building Regulations.

While carrying out the initial verification, you will be looking for non-compliant installation methods, as well as non-compliant equipment and materials. This can be anything from an incorrectly terminated protective equipotential bonding conductor (see Section 6 of this Guide) to equipment from overseas that does not meet the requirements of EU and British Standards.

The following checks should be carried out before testing:

(a) leads of test equipment are in good condition and fit for purpose;
(b) test instruments are calibrated, with a valid, current calibration certificate or proof of an ongoing accuracy check;
(c) probes should have no more than 4 mm of exposed tip (GS38 recommends 2 mm);
(d) test equipment is in good repair (remember that test equipment can be dropped and damaged even though it is within its calibration period);
(e) all necessary personal protective equipment (PPE) is in good condition;
(f) other people who may be affected by the testing are informed before the testing is carried out; and
(g) all electrical equipment is disconnected from the circuits that are being tested, including lamps.

9.2 Initial inspection

9.2.1 Why do you need an inspection?

Regulation 641.1 of the IET Wiring Regulations states that:

Every installation shall, during erection and on completion before being put into service, be inspected and tested to verify, so far as is reasonably practicable, that the requirements of BS 7671 have been met.

Regulation 641.4 continues:

> *Precautions shall be taken to avoid danger to persons and livestock, and to avoid damage to property and installed equipment, during inspection and testing.*

The IET Wiring Regulations require the following information about the installation that is to be tested to be made available to the person carrying out the testing procedures:

(a) maximum demand.

(b) number and type of live conductors (some commercial and most larger industrial premises have a three-phase supply, which consists of three or four live conductors. Smaller domestic premises usually have two, i.e. line and neutral).

(c) type of earthing arrangement (see Section 6 of this Guide).

(d) nominal voltage (in most supplies in the UK, 230 V or 400 V with an allowable tolerance; see Section 7 of this Guide).

(e) load current and supply frequency (see Section 3 of this Guide for frequency).

(f) prospective fault current at the origin (this is the highest short-circuit or earth fault current that can be expected at the incoming supply of the installation).

(g) external earth fault loop impedance (EFLI; with the measurement value of Z_e): this is the resistance of the line and earth conductor from the owner's installation to the distribution transformer.

(h) type and rating of overcurrent device.

9.2.2 Sequence of tests

The tests to be carried out, where relevant, during the initial verification of new installations are listed in Section 9 of the *On-Site Guide*.

Inspection and testing assessments carried out by training providers are considered practical assessments and the candidate will therefore be required to wear appropriate PPE. In most cases, safety boots that comply with the relevant standard are the minimum requirement.

9.3 The tests in detail

▼ **Table 9.1** Tests for initial verification

Test	Corresponding Regulation
Tests to be carried out before the power supply is connected (i.e. isolated)	
Continuity of protective conductors, including continuity of protective equipotential bonding (see Section 9.4.1) and continuity of circuit protective conductors (cpcs) (see Section 9.4.2).	Regulation 643.2.1
Continuity of ring final conductors (see Section 9.4.3).	Regulation 643.2.1
Insulation resistance (see Section 9.4.4).	Regulation 643.3
Polarity (by continuity method) (see Section 9.4.5).	Regulation 643.6
Tests to be carried out after the power supply has been connected (energized)	
Check polarity of supply using an approved voltage or polarity tester (see Section 9.5.1).	GS38
External EFLI (Z_e) (see Section 9.5.2) and EFLI (Z_s) (see Section 9.5.3).	Regulation 643.7.3
Prospective fault current (see Section 9.5.4).	Regulation 643.7.3.201
Phase sequence (see Section 9.5.5)	
Functional testing of equipment including residual current devices (RCDs) (see Section 9.5.6).	Regulation 643.10

Some things you will need:

(a) appropriate PPE;
(b) an insulated screwdriver set, including a torque screwdriver suitable for the application;

(c) suitable test equipment that complies with the guidance given in GS38, including:
   (i) a low-resistance ohmmeter;
   (ii) an insulation resistance tester;
   (iii) an EFLI tester;
   (iv) an RCD tester;
   (v) an earth electrode tester; and
   (vi) a polarity tester for three-phase power supplies;
(d) a lead to link conductors at the consumer unit or some other means of connecting cable together;
(e) a connector block; and
(f) a wander lead (sometimes called a 'long lead').

You will also need to consider how you are going to deduct the resistance of the test leads from your final result. The reason you need to do this is that you will be looking to identify the value of the circuit; the test leads are not part of the circuit and therefore their resistance should not be included in the final result. Reducing the resistance can be done by either:

(a) measuring the value of the resistance by connecting the test leads together and testing and recording their resistance so that it can be deducted from the end-result; or
(b) as most test instruments have a zero or null function (similar to the way you would zero a set of kitchen scales), carrying out the procedure as instructed by the manufacturer's guidance.

▼ **Figure 9.1** Nulling/zeroing the tester

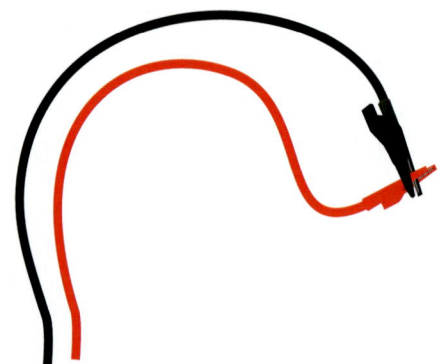

The following sequence of tests is to be carried out before the electrical supply is connected.

When nulling the test leads it is important that the crocodile clip jaws which are directly attached to the leads are placed together as shown in Figure 9.1. Failure to do so introduces resistance within the clips via the spring and its contact points and so recorded results will be inaccurate.

9.4 Dead tests

9.4.1 Continuity of protective equipotential bonding conductors

This test is to identify that the protective equipotential bonding conductors are continuous or unbroken and, subsequently, correctly connected.

The purpose of this test is to ensure that all exposed- and extraneous-conductive-parts will be at the same potential, hence the word 'equipotential' ('equal' + 'potential'). The test is carried out with one end of the bonding conductor disconnected, and this is usually easiest to achieve at the main earthing terminal (MET).

This test method can also be used to confirm a bonding connection between extraneous-conductive-parts where it is not possible to see a bonding connection, for example, where bonding clamps have been 'built-in'. The test would be done by connecting the leads of the test instrument between any two points, such as metallic pipes, and looking for a low reading of the order of 0.05 Ω; but do keep in mind that such a reading for a single-core 6.0 mm^2 or 10.0 mm^2 conductor equates to 15 m and 25 m, respectively.

NOTES:
1 It should be noted that not all low-resistance ohmmeters can read a value which is this low.
2 This is **not** the R_2 resistance measured from the MET to the bonding clamp or adjacent pipework earth.

For this test, you will need test equipment (a low-resistance ohmmeter) that complies with the guidance given in GS38.

To complete this test, you need to follow these simple steps in order.

ONE: isolate the supply (see the safe isolation procedure in Section 2 of this Guide and on the Student's Guide video series on IET.TV found here: https://tv.theiet.org/index.html).

TWO: disconnect one end of the conductor at the consumer unit.

THREE: measure the resistance of the test leads or, better still, null/set them to zero. If a wander lead is required, the resistance of this lead must also be nulled or deducted, following the test.

FOUR: connect one test lead to the disconnected conductor at the MET, which may be in the consumer unit.

FIVE: connect the other lead to the opposite end of the conductor (depending on the distance between the ends of the bonding conductor, you may be required to use a wander lead).

SIX: if you haven't nulled the test instrument, remember to subtract the resistance of the leads, including the wander lead, from the total resistance shown on the meter. This will give you the resistance of the bonding conductor.

SEVEN: ensure that the bonding conductor is reconnected on completion of the test (if this is not done, the installation will not be bonded).

While carrying out this test, a visual inspection can also be done to ensure that the correct type of earthing clamp is present, complete with the correct label, and that the conductor is one continuous length (has not been broken and reconnected) where it is bonding more than one service or extraneous-conductive-part.

9.4.2 Continuity of the cpc

For reasons of safety, the cpc is one of the most important conductors in the installation. If something goes wrong at any point, such as a line-to-earth fault, the cpc will carry sufficient fault current around the EFLI path to operate the protective device.

As the person doing the tests, you will need to verify that the cpcs of the radial circuits are in place and are connected throughout the circuits that you are testing in the installation.

The instrument you will use to carry out this test is a low-resistance ohmmeter (sometimes called a 'continuity tester'), which should be set on the lowest scale for measurement.

This test is commonly referred to as a 'dead test', which means that it must be carried out on an isolated circuit. Continuity of the cpc can be determined using one of two methods. When carrying out either of these tests on a circuit with, for example, light switches, it will be necessary to close the switches in order to test all parts of the circuit to obtain an accurate measurement.

Continuity test method 1 (also known as the 'link method')

On the IET model Schedule of Test Results form, there are two permissible values that may be recorded when verifying the continuity of the cpc. These are $R_1 + R_2$, or just R_2. R_1 represents the line conductor and R_2 represents the cpc. Continuity test method 1 (Method 1) essentially uses the line conductor as an extended test lead. It runs alongside the cpc, so it might as well be used to help obtain the continuity reading for the cpc.

For this test, you will need:

(a) test equipment (a low-resistance ohmmeter) that complies with the guidance given in GS38;
(b) appropriate PPE; and
(c) a temporary link – a short length of cable or a connector so that the line and the cpc can be joined together at one end of the circuit, usually at the consumer unit.

To complete this test, you need to follow these steps in order.

ONE: the supply will be isolated (see Section 2 of this Guide for information about the safe isolation procedure).

TWO: measure the resistance of the test leads or, better still, null/set the tester to zero.

THREE: using a short length of cable or a connector block, connect the line and the cpc together at the consumer unit.

FOUR: the test instrument should be set to a low resistance setting and the probes should be applied to the cpc and line conductor at each accessible point of the circuit, i.e. at the switch and ceiling rose. During this test, current will flow from one end of the test instrument, for example, from the red lead, through the line conductor of the circuit to the consumer unit, then through the temporary link back through the cpc to the test instrument, at which point the resistance will be measured and displayed on the test instrument.

▼ **Figure 9.2** Example of Method 1 on a radial final circuit

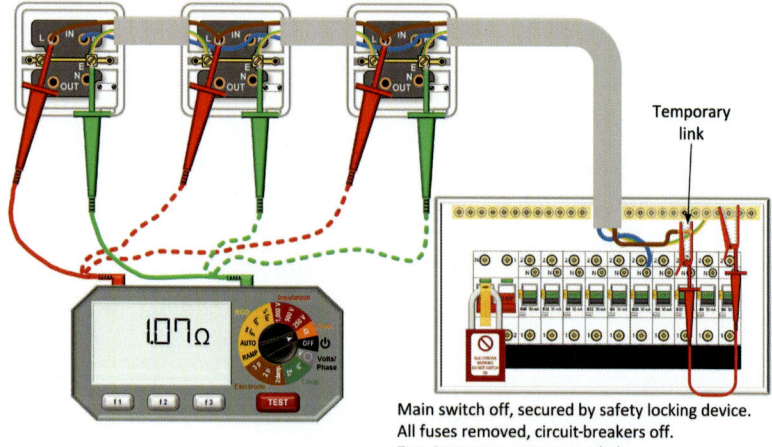

Main switch off, secured by safety locking device.
All fuses removed, circuit-breakers off.
Zero instrument across test link.
NOTE: remember to remove temporary link after test.

FIVE: the highest reading obtained for each circuit should be recorded on the Schedule of Test Results as ($R_1 + R_2$). Note that the readings should increase as you move further away from the consumer unit.

The result obtained using this test method can be added to the external EFLI to calculate the total resistance of the EFLI path (measurement value Z_s).

Continuity test method 2 (also known as 'long lead' or 'wander lead')

This method (Method 2) will prove that the cpc is continuous. This method is usually only used where the circuit is wired in steel enclosures (trunking/conduit/ducting) and where parallel paths to the cpc may be present. It is also useful if there are many enclosed metal fittings, where dismantling them would be impractical.

You will need:

- **(a)** a low-resistance ohmmeter that complies with the guidance given in GS38;
- **(b)** appropriate PPE; and
- **(c)** a wander lead (available to purchase at most electrical wholesalers).

There is no need to carry out this testing method if Method 1 has already been carried out.

To complete this test, you will need to follow these simple steps in order.

ONE: isolate the supply (see Section 2 of this Guide for information on the safe isolation procedure).

TWO: measure the resistance of the test leads or, better still, null or make them zero.

THREE: either connect the wander lead directly to the test instrument or connect the lead from the test instrument to one end of the wander lead. The wander lead from the test instrument can be connected to the earth terminal at the consumer unit. Measurements shall be taken from all accessible points on the circuit and the highest reading obtained must be recorded and written on the Schedule of Test Results as R_2.

▼ **Figure 9.3** Example of Method 2 on a radial circuit

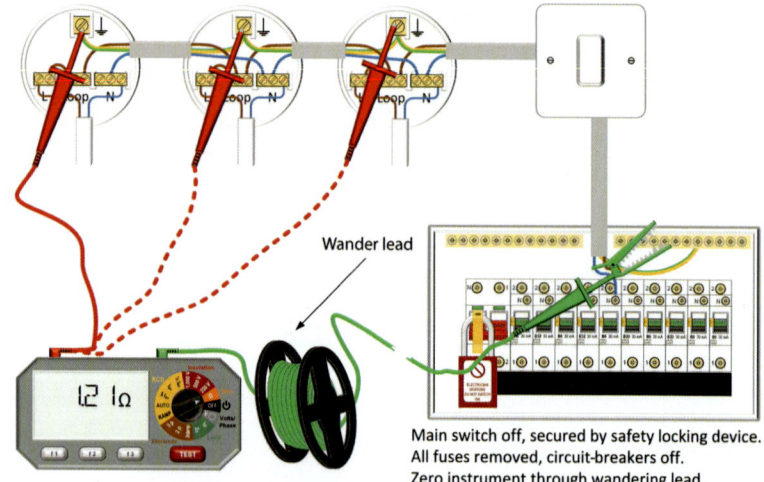

Main switch off, secured by safety locking device.
All fuses removed, circuit-breakers off.
Zero instrument through wandering lead.

FOUR: take the tester and hold the short lead test probe against the earth terminals at each accessible piece of equipment and accessory at its point of connection to the circuit you are testing, and test it. This ensures that you will identify the furthest point on the circuit which will produce the highest reading.

FIVE: this highest reading is to be recorded on the Schedule of Test Results as R_2.

9.4.3 Continuity of ring final circuit conductors, including protective conductors

The purpose of this test is to confirm that:

(a) the conductors form a complete ring. When conductors are not connected to complete a full ring, various parts of the circuit may have more current passing through them than the cable is designed to handle. This may lead to overheating and fire.

(b) there are no interconnections. Interconnections suggest that the circuit may have been connected in such a way that there are bridging cables between two parts of the circuit. This can create additional strain on certain parts of the circuit and can also make it difficult to verify other test results.

(c) the polarity is correct in all socket-outlets. This test identifies that the line conductor is connected to the line terminal of the socket-outlet and the neutral is connected to the neutral terminal of the socket-outlet.

(d) there will be an increased risk of electric shock if the polarity is incorrect, for example, if a lamp that has an Edison screw (ES)-type fitting is plugged into a socket with incorrect polarity.

When this test is carried out correctly, it also gives you the $R_1 + R_2$ value of the ring final circuit and can be used to identify any spurs in the circuit.

The instrument that you will use to carry out this test is, once again, the low-resistance ohmmeter and, as before, this should be set to the lowest value possible.

You will need:

(a) a low-resistance ohmmeter that complies with the guidance given in GS38;
(b) appropriate PPE;
(c) three connectors; and
(d) a test adapter which plugs into the socket-outlets. (This means that test steps seven, eight and nine can be done with the socket-outlets fixed in place.)

NOTE: Some instrument manufacturers supply an instrument lead with a BS 1363 plug for carrying out testing at socket-outlets.

▼ **Figure 9.4** Typical socket-outlet test adapter (Image provided courtesy of Socket and See)

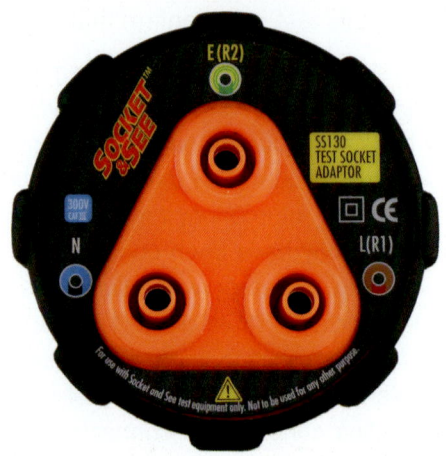

To complete this test, you need to follow these simple steps in order.

ONE: isolate the supply (see Section 2 of this Guide for information on the safe isolation procedure).

TWO: measure the resistance of the test leads or, better still, null or zero the test instrument.

THREE: identify each leg of the ring (incoming and outgoing); this is usually done at the installation stage by marking the cables with either **in** and **out** or **1** and **2**, etc.

FOUR: test between the line conductors of the ring. Record this value on the Schedule of Test Results (r_1).

9

▼ **Figure 9.5** Connections for ring final circuit continuity testing: step 1

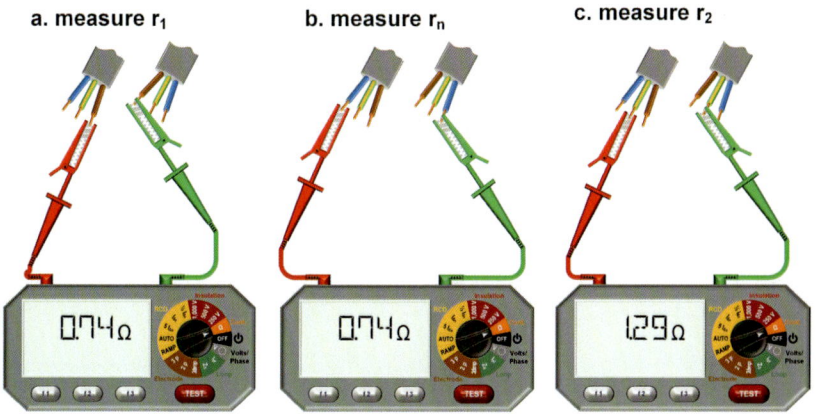

a. measure r_1 b. measure r_n c. measure r_2

FIVE: test between the neutral conductors of the ring. Record this value on the Schedule of Test Results (r_n). This value should be approximately the same as the line conductor.

SIX: test between the ends of the cpc of the ring. Again, record this value on the Schedule of Test Results (r_2). If the conductor sizes are smaller than the phase and neutrals, this value will be higher.

SEVEN: securely join together, with a connector block, the incoming line to the outgoing neutral conductor and the outgoing line to the incoming neutral conductor, as shown in Figure 9.9.

Figure 9.6 Line from leg one connected to neutral from leg two

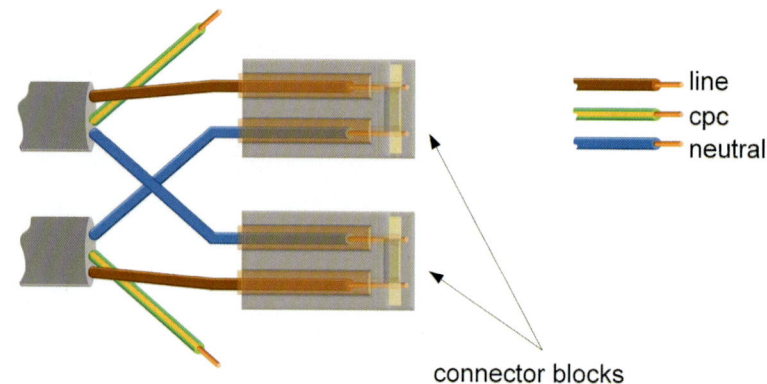

connector blocks

EIGHT: test between line and neutral at each socket-outlet on the circuit using the socket-outlet test adapter. The resistance should be almost the same at each socket-outlet. A higher reading should be investigated, as it may be a spur on the circuit, which should be recorded on the Schedule of Test Results or it may be as a result of a loose connection.

NINE: disconnect the neutral conductors from the connector block and connect the cpc and line conductors as shown in Figure 9.10. Repeat the test at the line and cpc terminals at each socket-outlet. Operate each switch on the socket-outlets to confirm polarity. Where the cpc has a different cross-sectional area (csa) to the line conductors, for example, where flat twin-and-earth cable is used, the resistance will increase as the tests move around the ring from the origin of the circuit, to a maximum of approximately $(r_1 + r_2)/4$ at the mid-point of the ring, and decrease as the test point moves back towards the origin of the circuit. Information on the expected maximum and minimum values can be found in Table 2.4 of IET Guidance Note 3.

▼ **Figure 9.7** Line from leg one connected to cpc from leg two

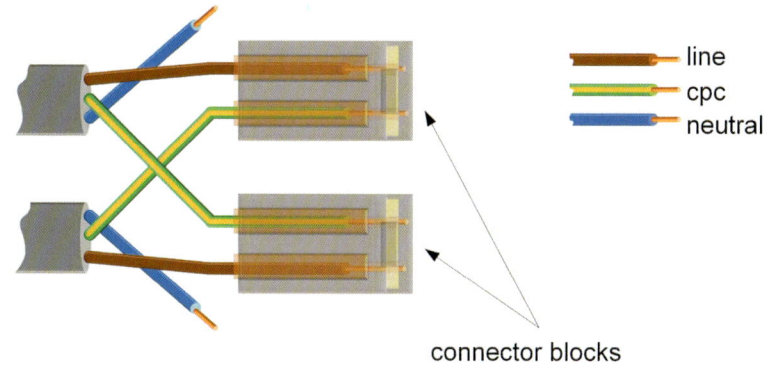

connector blocks

9.4.4 Insulation resistance (Regulation 643.3)

Important: this test can cause damage to electrical and electronic equipment if the procedures are not carried out correctly. There is also a risk of electric shock from the direct current (DC) produced by the test instrument.

This test is carried out to determine that the resistance of the insulation around the conductors is sufficient and that there are no short circuits between the live conductors or between the live conductors and the cpc. The insulation is designed to stop current escaping from the conductor. However, cable insulation can deteriorate with age and might also be weakened if, for example, cables are crushed beneath floorboards, damaged by clips or suffer damage due to being pulled through holes in joists, etc.

This test can be carried out on a complete installation or on a single circuit – whichever is suitable or required. It is necessary to perform this test in order to find out if there is likely to be any leakage of current through the insulated parts of the installation.

The insulation resistance shall be measured between:

(a) live conductors, and
(b) live conductors and the protective conductor connected to the earthing arrangement. During this measurement, line and neutral conductors may be connected together.

The instrument used to carry out this test is an insulation resistance tester. This method works a little like a pressure test. It subjects the conductors to a higher DC voltage for a short time, to ensure that the insulation surrounding them contains the voltage. If the insulation is not sufficient, there is a risk of a short circuit or a line-to-earth fault. Table 9.2 shows the required setting for the test instrument in relation to the voltage of the circuit.

Where connected equipment is likely to influence the measurement or the result of the test, or be damaged, the test shall be applied prior to the connection of such equipment, in accordance with Table 9.2. Following connection of the equipment, a test at 250 V DC shall be applied between live conductors connected together and the protective conductor connected to the earthing arrangement. The insulation resistance shall have a value of at least 1 MΩ.

▼ **Table 9.2** Minimum values of insulation resistance (From Table 64 of the IET Wiring Regulations)

Nominal voltage of the circuit	DC test voltage to be applied (V)	Minimum insulation resistance (MΩ)
0 V – 50 V (SELV and PELV)	250	0.5
Up to and including 500 V, with the exception of SELV and PELV, but including FELV	500	1.0
Above 500 V	1,000	1.0

Although an insulation resistance value of not less than 1.0 MΩ for the whole installation complies with the IET Wiring Regulations (the resistance is usually much higher), where the measured insulation resistance is less than 20 MΩ, each circuit should be tested separately. This will help to identity, firstly, which circuit has a lower value and, secondly, a possible defect in the insulation of the conductor.

Before carrying out the test, ensure that all:
- **(a)** lamps are removed;
- **(b)** electronic devices are removed;
- **(c)** plug tops are removed from the socket-outlets; and
- **(d)** protective devices are in the 'on' position and the main switch is in the off position.

To complete this test, you need to follow the steps below in order.

ONE: set the insulation resistance tester to the correct voltage needed for the circuit to be tested (500 V for 230 V circuits).

TWO: push the test button with the test leads disconnected. The resistance shown on the screen should be the high Ω reading, for example, >999 MΩ.

THREE: join the test leads together and operate the test button again. The resistance shown on the screen should be very close to 0.0 Ω. Steps two and three identify that the test instrument is working.

FOUR: the quickest way to identify that the insulation resistance is sufficient on all circuits is to test them all at the same time. When testing lighting circuits that have two-way or intermediate switching, each switch must be operated and the circuit re-tested after each change of switch position, so that all conductors, including strappers, have been tested. The terminals to which the test probes need to be applied are shown as follows.

FIVE: record the test results on the Schedule of Test Results.

SIX: use Table 9.2 of this Guide or Table 10.3.3 in the *On-Site Guide* to verify that the results comply with the IET Wiring Regulations.

▼ **Figure 9.8** Testing insulation resistance between line and neutral

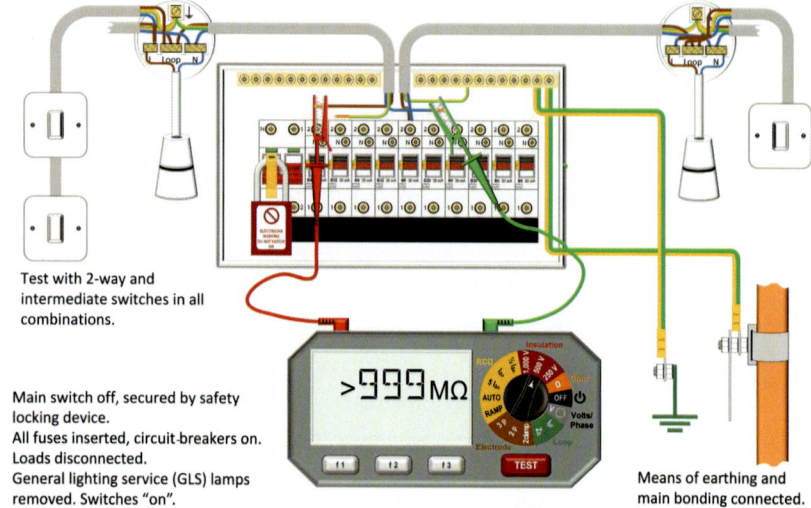

Test with 2-way and intermediate switches in all combinations.

Main switch off, secured by safety locking device.
All fuses inserted, circuit-breakers on.
Loads disconnected.
General lighting service (GLS) lamps removed. Switches "on".

Means of earthing and main bonding connected.

▼ **Figure 9.9** Testing insulation resistance between neutral and earth

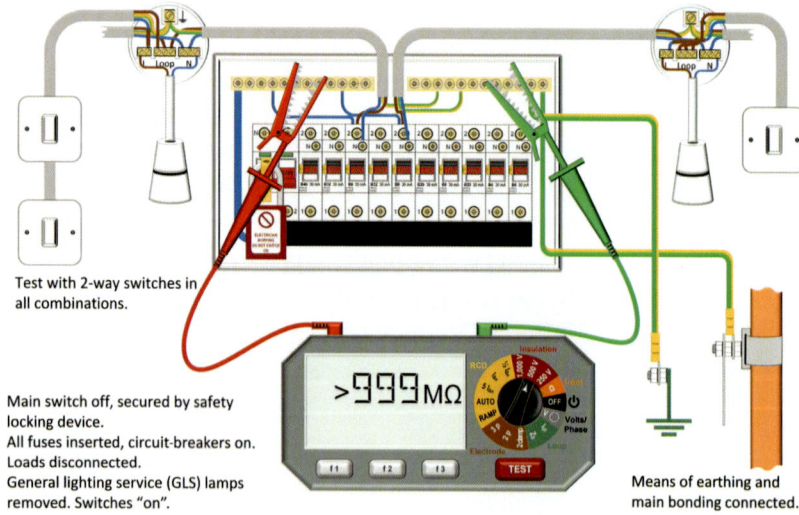

Test with 2-way switches in all combinations.

Main switch off, secured by safety locking device.
All fuses inserted, circuit-breakers on.
Loads disconnected.
General lighting service (GLS) lamps removed. Switches "on".

Means of earthing and main bonding connected.

▼ **Figure 9.10** Testing insulation resistance between line and earth

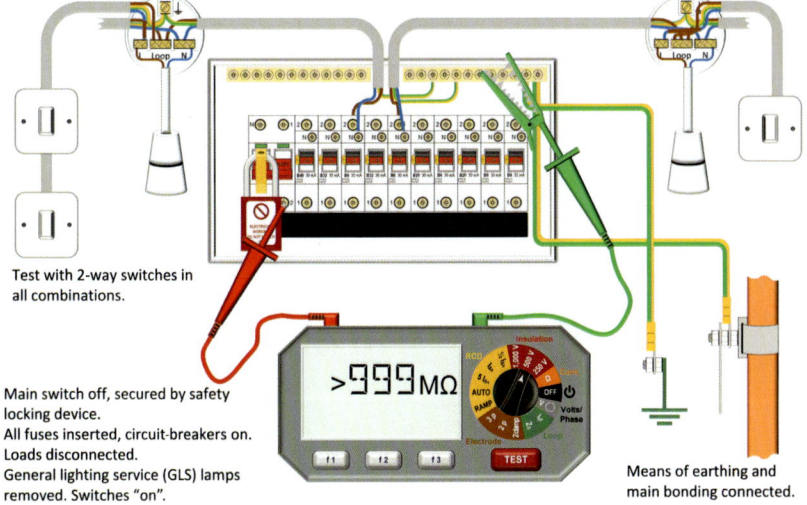

9.4.5 Polarity

This test is to confirm that all:

(a) single-pole protective devices are connected to the line conductors only;
(b) single-pole devices are also connected to the line conductors only;
(c) ES lampholders have the centre pin connected to the line conductor; and
(d) accessories are correctly connected.

In most cases, polarity can be confirmed by operating the switch at the same time as the continuity of the cpc test is carried out. When the switch is operated, the reading from the low-resistance ohmmeter will indicate a low resistance when the switch is closed and a high resistance when the switch is open. Polarity can also be checked visually by verifying that the cores have the correct colour identification at their terminals. Where the Method 2 test is used to confirm continuity of the cpc, then polarity will need to be confirmed by this test.

Using the low-resistance ohmmeter with the lowest value of Ω possible, complete this test by following these simple steps in order.

▼ **Figure 9.11** Testing polarity using Test Method 1

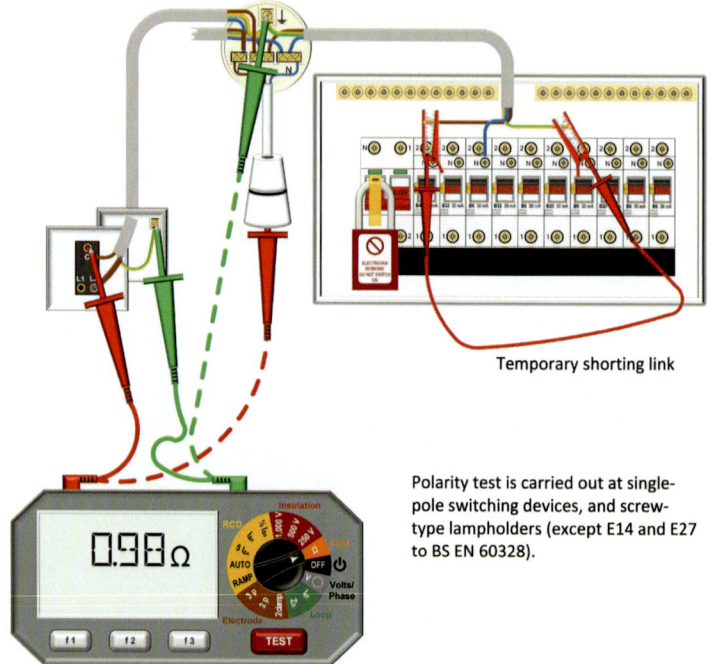

Temporary shorting link

Polarity test is carried out at single-pole switching devices, and screw-type lampholders (except E14 and E27 to BS EN 60328).

ONE: isolate the supply (see Section 2 of this Guide for information on the safe isolation procedure).

TWO: put the main switch in the 'off' position, remove all fuses or put protective devices in the 'off' position, and remove all lamps and equipment.

THREE: using a short lead with a connector block on each end, connect the line and cpc of the circuits to be tested together at the consumer unit.

FOUR: measure the resistance of the test leads or, better still, null or zero the test instrument.

FIVE: at each point of each circuit, test between the line and the cpc (either at the switch or at the lampholder).

SIX: close the switch that is controlling the light; the test instrument should read a low resistance. Opening the switch will then result in a high resistance. This confirms that the line conductor has been switched and therefore confirms polarity.

The following tests are live tests and should only be carried out under the supervision of an electrically skilled person, such as a lecturer or an electrician.

9.5 Live tests

9.5.1 Polarity

This test confirms that there is a voltage present when you switch on the supply. The test is carried out at the line and neutral terminals of the incoming supply to the consumer unit.

The instrument you use must comply with GS38 and be an approved voltage indicator.

To complete this test, follow these steps in order.

ONE: place the test probes of the voltage indicator onto the neutral and line terminals of the incoming supply at the main switch. Your device should indicate a live supply.

TWO: place the test probes of the voltage indicator onto the earth and line terminals of the incoming supply at the main switch. Your device should indicate a live supply.

THREE: place the test probes of the voltage indicator onto the earth and neutral terminals of the incoming supply at the main switch. Your device should indicate NO supply.

Polarity will also be confirmed when carrying out the test to measure external EFLI using a three-lead test instrument.

9.5.2 Earth fault loop impedance (Z_e) (Regulation 643.7.3)

Z_e is a measurement of the external EFLI of the installation.

To carry out this test, the installation should be isolated from the supply and the main earth should be disconnected from the MET. This is to avoid any parallel paths through any other earthed metalwork within the building.

9

The instrument used for this test is an EFLI tester. The measured value for Z_e can be found by completing the following steps in order.

ONE: acquire the necessary PPE and insulated tools.

TWO: isolate the supply to the consumer installation (the incoming supply must remain live to carry out this test).

THREE: disconnect the earthing conductor (this is essential to avoid obtaining readings from parallel earth paths).

FOUR: set your instrument to measure Z_e (you may have to refer to the manufacturer's instructions for the test instrument to find out what the correct setting is).

FIVE: if using a two-lead instrument, one lead should be connected to the disconnected incoming main earth conductor and the other lead to the incoming line conductor on the supply side of the main switch.

SIX: press the test button and record the result on the Electrical Installation Certificate. Where the consumer unit is at the origin it will be recorded as Z_{db} on the Schedule of Test Results (the incoming supply should remain live).

SEVEN reconnect the incoming main earth conductor. This is a very important step.

▼ **Figure 9.12** EFLI two-lead test

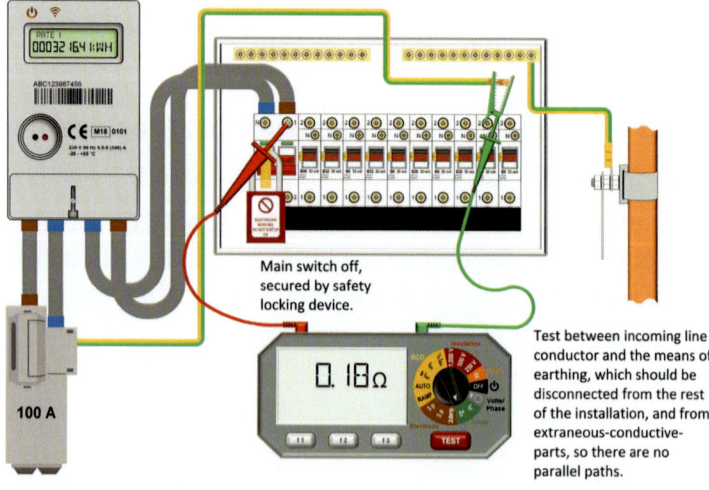

NOTE: When undertaking your practical examination and being assessed, if you do not reconnect the earth, you will fail the entire section of this exam. If you were to forget this step while working on an installation, you would essentially have an un-earthed system, which would present a serious risk of electric shock.

Remember: This is a LIVE test and care should be taken at all times during the test procedure.

If using a three-lead instrument, the manufacturer's instructions should be followed. Where a neutral is available at the point of test, the general requirement is for the connections to be made to the incoming line, neutral and the disconnected earthing conductor.

9.5.3 EFLI of each circuit (Z_s)

Z_s is the measurement of the earth loop impedance of a final circuit, including the supply conductor. This test is carried out to ensure that the required disconnection times of each circuit in the installation will be achieved if a fault to earth occurs.

The tests are carried out at the furthest point of each final circuit within the installation. If the measured values of Z_s for each circuit are lower than or equal to the maximum values given in Appendix B of the *On-Site Guide* (where adjustments for temperature have been made), the required disconnection times will be achieved. See Appendix B of the *On-Site Guide* for more information.

To complete this test, follow these steps in order.

- **ONE:** this is a live test, so the circuit should be energized at the time of testing.
- **TWO:** prepare the circuit and the test instrument so that each available point can be tested (special adapters are available so that instruments can be plugged into socket-outlets and lampholders, and some test instrument manufacturers supply these with the test instrument).
- **THREE:** in accordance with the manufacturer's instructions for the test instrument, test each accessible point of the circuit and record each reading that you obtain.
- **FOUR:** verify that the highest result obtained is within the required values given in the IET Wiring Regulations and record it on the Schedule of Test Results.

9.5.4 Prospective fault current (I_{pf}) test

Regulation 643.7.3.201 requires that the I_{pf} under both short-circuit and earth fault conditions is measured, calculated or determined by another method, such as enquiry.

An I_{pf} test is carried out to find the maximum current that could flow between line and neutral, between line and line and between line conductors and earth. An I_{pf} test instrument is normally a function of an EFLI tester and the result is measured in kA. Regulation 434.1 of the IET Wiring Regulations requires that the prospective fault current should be determined at every relevant point of the installation, i.e. at every point where a protective device is required to operate under fault conditions, including at the origin of the installation. If the I_{pf} is measured at the consumer unit and is found to be satisfactory, it can be assumed that all of the circuits supplied from that consumer unit will also be satisfactory.

This value can be obtained in three ways:

(a) by enquiry to the supplier;
(b) by calculation; or
(c) by measurement.

Measurement of the I_{pf} can be obtained by using a similar method of testing to that used to determine Z_e; however, the incoming earth should be connected when carrying out this test so that any parallel paths are included. The test equipment manufacturer's instructions should be followed so that the correct results can be obtained.

9.5.5 Phase sequence (Regulation 643.9)

In the case of three-phase installations, you should check that the phase sequencing is correct. This can be done by carrying out a simple test, which requires a phase sequence tester. The tester will have three leads, generally identified as brown, black and grey, or L1, L2 and L3, respectively. In some older installations, you will come across conductors identified with red, yellow and blue: the order of sequence is the same. The leads should be held on to the corresponding terminals – the rest is automatic. The test instrument will display a reading, similar to that shown in Figure 9.16.

▼ **Figure 9.13** Typical readings displayed on a phase sequence tester

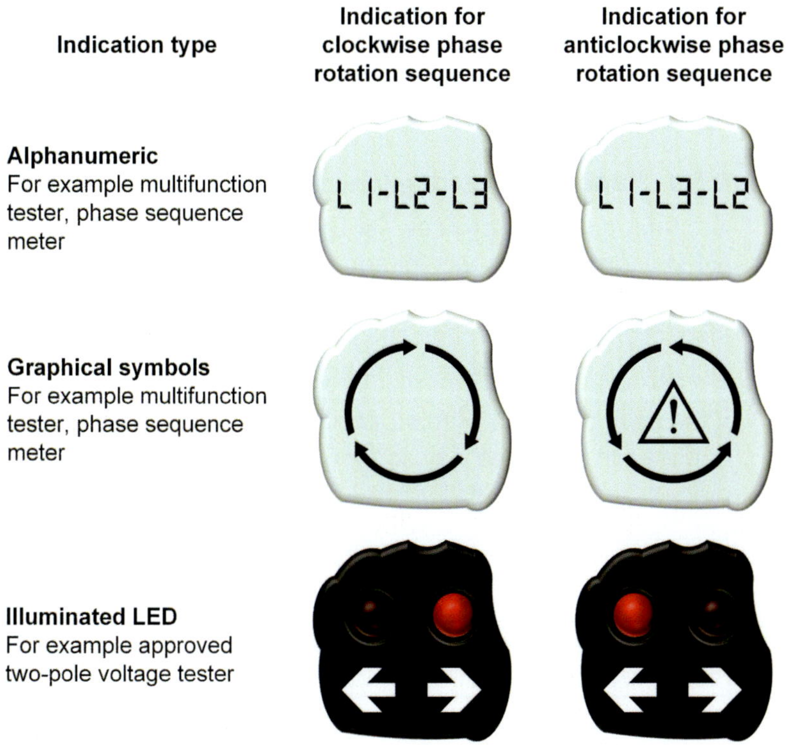

Indication type	Indication for clockwise phase rotation sequence	Indication for anticlockwise phase rotation sequence
Alphanumeric For example multifunction tester, phase sequence meter	L1-L2-L3	L1-L3-L2
Graphical symbols For example multifunction tester, phase sequence meter	(clockwise circular arrow)	(anticlockwise circular arrow with warning triangle)
Illuminated LED For example approved two-pole voltage tester	(red LED lit on right, arrows ← →)	(red LED lit on left, arrows ← →)

9.5.6 Functional testing of RCDs

The instrument used to carry out this test is an RCD tester: it measures the time it takes for the RCD to interrupt the supply of current flowing through it, with the value of measurement given in either seconds or milliseconds. RCD test instruments generally come supplied with leads that have plug tops for testing socket-outlet circuits and a combination of bayonet and ES lamp attachments for testing lighting circuits, as well as the standard leads for testing other circuits. Most modern RCD test instruments have an automatic function that carries out the correct sequence of tests, only requiring the person carrying out the test to close the RCD after each operation.

30 mA RCD with integral test button

It has been found in the past that if an RCD has not been operated for some time, this can have an effect on the time it takes for the device to operate. To avoid this, and to prevent failure of mechanical operation, the test button must be operated every six months.

9.5.7 Functional testing/checking of AFDDs and SPDs

Arc fault detection devices (AFDDs) and surge protective devices (SPDs) are requirements for some installations, and these generally are autonomous in their operation. Should these be fitted, then inspection of their condition and, where a built-in test facility is included, a test of operation is required.

If the device is fitted with a test button or switch, then this should be operated in the same way as for RCDs.

Many of these devices will have a status indicator which should be checked to confirm that the device is operational and in a suitable condition.

	Test your knowledge
1.	What can be identified by carrying out an insulation resistance test?
2.	When testing for the continuity of a cpc in a circuit, should the highest or the lowest reading obtained be recorded on the Schedule of Test Results?
3.	Which result should be recorded on the Schedule of Test Results following an insulation resistance test of a new circuit: the highest or the lowest reading?
4.	What needs to be done with the line and earth conductors when using Method 1 to identify continuity of the cpc?
5.	Where can maximum permissible measured values of Z_S be found in the *On-Site Guide*?
6.	When testing for continuity on a lighting circuit, you cannot obtain a low Ω reading. What else might you have to do to test the circuit correctly?
7.	According to GS38, what is the recommended length in mm that the tips of the test probes should be?

Fault Finding   10

This Section provides information on the following topics:
- ▶ What is a fault?
- ▶ Identifying a fault
- ▶ Fault location
- ▶ Fault rectification

10.1 What is a fault?

Before fault finding can begin, you must know what it is that you are trying to find.

A fault can be a combination of various elements, or a single situation, that prevents an electrical circuit from functioning correctly.

There are generally three main types of fault that can occur on an electrical circuit:

(a) **short circuit:** where the length or resistance of a circuit is reduced (shortened), for example, by a screw or a nail through a cable creating continuity between live conductors, resulting in an overcurrent;

(b) **open circuit:** where there is no complete path for current to flow, for example, where cable has been removed from the terminal or broken completely; and

(c) **earth fault:** where a fault current flows from a live conductor to earth.

The following examples are not defined as faults in Part 2 of the IET Wiring Regulations; however, awarding bodies include these in exams to assess fault identification and location techniques:

(a) **high resistance:** where the flow of current is restricted, for example, due to a loose connection or a stranded cable not terminated correctly;

(b) **cross polarity:** where the line and neutral conductors have been connected the wrong way round; and

(c) **interconnection of conductors:** where conductors have been connected into the wrong terminals – for example, on a heating control unit where the pump and thermostat have been connected the wrong way round.

Faults and other problems can occur for a number of reasons. Some examples are as follows.

10.1.1 Short circuit

This is a common fault normally caused by some form of mechanical damage to a cable, such as a screw or nail being driven through the conductors or the cable being crushed by furniture or a heavy object. This type of fault will, in most cases, operate the protective device immediately, whether it blows a fuse or trips a circuit-breaker.

10.1.2 Open circuit/high resistance

An open circuit is commonly the result of a break in the circuit due to some form of mechanical damage, a broken conductor, the failure of a component as it has become old and no longer suitable for continued use, or a conductor that has simply become loose from its terminal. This will normally result in the equipment not operating at all.

10.1.3 Cross polarity

To put it simply, cross polarity is where the conductors are connected the wrong way round, such as the line connected into the neutral terminal and the neutral connected into the line terminal. This can happen where a job has been rushed or where there has been insufficient light to correctly identify cable terminals when installing. This is not acceptable and will be identified when the correct test procedures are carried out prior to an installation being energized.

10.1.4 Interconnection of conductors

When installing control systems, it can sometimes be difficult to keep track of which conductors you are connecting where. If the cables have not been clearly identified, it is possible to get them mixed up. A good example is a heating control system that consists of a number of component parts, such as thermostats, pumps and valves. All the conductors from these parts come back to the main terminal box at the controller. If the conductors have not been identified correctly, they may end up in the wrong terminals, or, if some maintenance work has been carried out, they may have been reconnected in the wrong place.

10.1.5 High resistance

High resistance can put additional strain on cables and potentially lead to hazardous situations, such as overheating and fires. This can be a result of damaged electrical equipment, loose connections (high resistive joints), etc.

If a conductor is forced into a terminal that is too small, it may damage the terminal, which will potentially result in high resistance.

If a conductor is connected into a terminal that is too large, there is a higher risk of the conductor not being sufficiently clamped and it may come loose. All components should be carefully selected so that they serve their intended purpose.

Figure 10.1 shows cable being terminated into connector blocks that are the wrong size. This could result in the cable not being held in the terminal correctly (if you try to use a conductor that is too small) or a connector block being damaged (if you try to use a conductor that is too large).

▼ **Figure 10.1** Incorrect size terminations or connector blocks

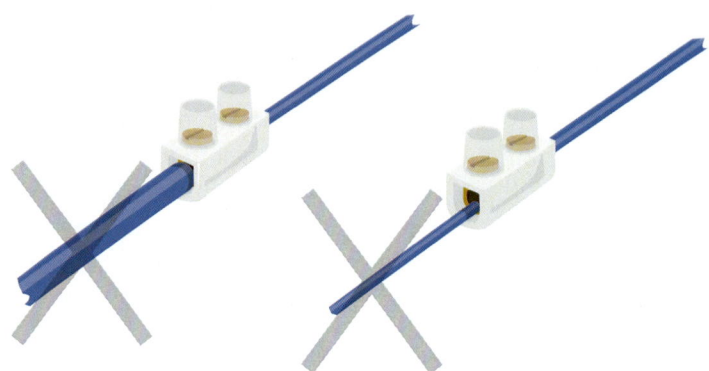

10.1.6 Overload

An overload is not a fault – it can be a result of incorrect design, the addition of too much current-using equipment, faulty equipment, etc. If the overload current is small, it may take some time for the protective device to operate, which can make it a difficult fault to identify and locate. One method of testing for overload is to use a clamp meter. This type of meter can be connected around a conductor to take a reading of the current that is flowing through it. This reading is used in conjunction with the current-carrying capacity of the cable; the rating of the equipment will help to identify if the circuit and any connected items are functioning correctly or not.

10.2 How to diagnose and locate a fault

Even before you take any test equipment from your van, there is a good chance you can identify the cause of the fault.

Step 1: communication

Communication is key. Speak to the customer to gather as much information as you can. You will need to prompt discussion: remember, not everyone is an electrician, and they may have done something that can affect the electrical installation without realising it. Hanging a picture, moving some furniture or having had a bathroom fitted recently can all be causes of faults. Alternatively, with changing seasons, the customer may have forgotten about an appliance they only use at certain times of the year; for example, an electric heater may have had a build-up of dust and debris, which, when switched on, has caused damage to the heating elements.

Once you have communicated with the customer and assessed the information available, you can begin further investigation.

Step 2: visual inspection

After communicating with all relevant people about the electrical fault that has been reported, a visual inspection should be carried out.

All electrical appliances and equipment should be removed and disconnected from the installation. This allows the electrician to identify whether it is the electrical installation, an appliance or a piece of equipment that is causing the problem.

Most faults are caused by some kind of mechanical impact, such as a nail through cable or damage to electrical equipment that has been moved or replaced, or by items of equipment that have reached the end of their functional life.

10.2.1 What to look out for

    (a) signs of general wear and tear;
    (b) cable not adequately contained or supported;
    (c) blackening around terminals and other electrical connection points;
    (d) cracked or damaged appliances;
    (e) water ingress;
    (f) corrosion;
    (g) incorrect equipment installed on circuits; and
    (h) signs of new construction or installation of electrical and non-electrical equipment, especially wall-mounted pictures, shelves, etc.

Step 3: identify the type of fault

In order to carry out fault finding safely and to identify the fault, the safe isolation procedure must be carried out (see Section 2). Once this has been done, you can test the circuit. Remember that the only part of the circuit to be tested is the wiring system, so all appliances must be disconnected and lamps removed from fittings.

The following are the types of fault you are most likely to encounter.

10.2.2 Short-circuit/earth fault

This fault occurs when live conductors of the same circuit are touching or have been connected by a conductive material, either to each other or to the protective conductor – commonly, a nail or screw through a cable. A live conductor to earth fault can be produced when cables are crushed against the earthed metal accessory boxes when fixing the accessory to the box. Care during installation can avoid this type of cable fault.

The first of the tests can be carried out with a low-resistance ohmmeter. Check for a short circuit: hold the probes of the tester across L-N, L-E and N-E. In an undamaged circuit, there should be a high resistance across these conductors.

If the result of the test indicates a low resistance between L-N, there is a short circuit. If it indicates a low resistance between L-E or N-E, there is a line to earth or a neutral to earth fault.

▼ **Figure 10.2** Nail through cable

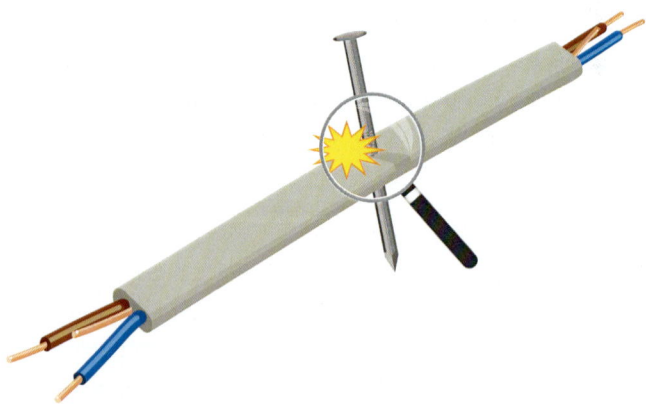

In Figure 10.2, a nail has gone through a cable, damaging the outer sheathing as well as the insulation surrounding the line conductor. As nails are made of a conductive material, electrical current is able to flow from the live conductor through the nail to the bare protective conductor, thus creating a path for an earth fault.

10.2.3 Open circuit

▼ **Figure 10.3** Line conductor damaged

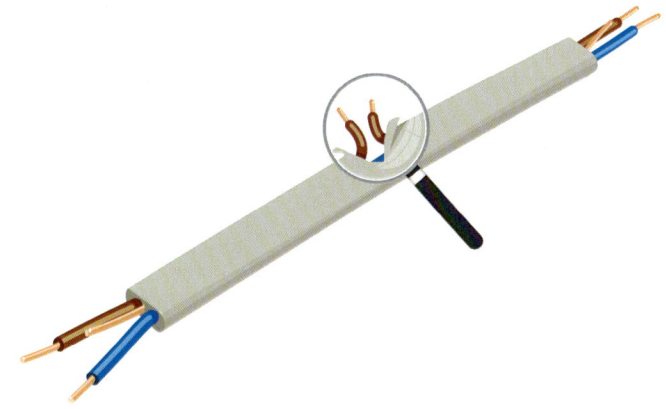

This can be a difficult problem to diagnose, especially on a lighting circuit or on circuits with switches, isolators and online starters. One method that can be used is a test of the continuity between two conductors, as when testing the continuity of the circuit protective conductor (cpc) on a final circuit. For this test, you will need to join the line and the cpc together at the consumer unit, locate the end of the circuit and carry out a low ohms test on all conductors. If the circuit has switches and/or isolators, you will have to carry out this procedure while operating all switches. This test needs to be carried out on each conductor combination, for example, L-E, L-N, and N-E. If a low ohms reading cannot be achieved for each test, this suggests that there may be an open-circuit fault.

10.2.4 High resistance

▼ **Figure 10.4** Poor termination, frayed strands of conductor

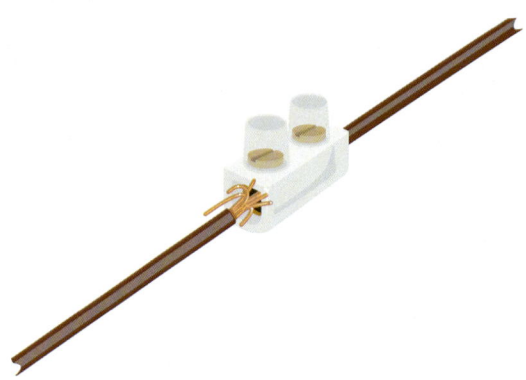

High resistance can be the result of a deteriorated termination, a poor connection or simply a faulty product or poor installation. This type of fault can potentially cause a fire and may go unnoticed for many years. The best method of avoiding this type of fault is to check and tighten all electrical connections to ensure they are nice and secure. You can test for this type of fault using a low ohms test instrument: you may find the resistance value higher than expected. Indications of suitable resistance values can be found in the maximum Z_s values in Table I1 in Appendix I of the *On-Site Guide*.

10.2.5 Cross polarity

▼ **Figure 10.5** Conductors incorrectly terminated

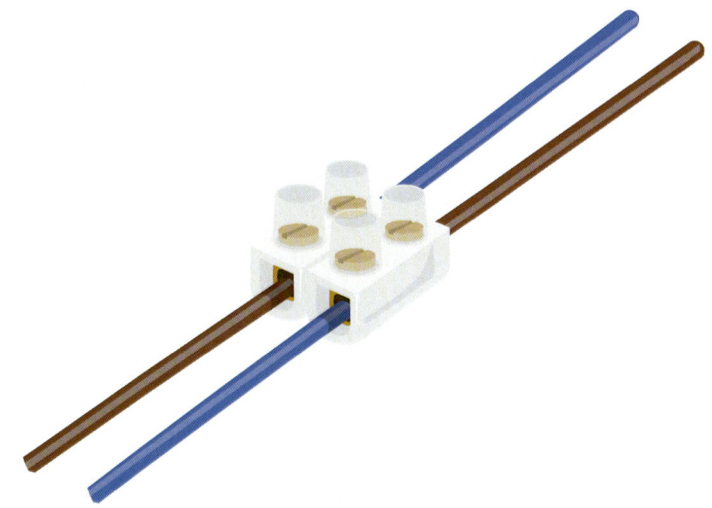

Cross polarity can be identified fairly easily. When a circuit has been connected incorrectly, with the live conductor connected to the neutral terminal of an accessory or piece of equipment, it can usually be identified visually. If this is not possible, an R1 + R2 type test can be carried out to identify if the conductors are in the correct location. If the L and E are connected together at one end of the circuit, you should be able to obtain a low ohms value when testing across the same conductors at the other end of the circuit. If a high reading is measured, check N to E; if a low reading is obtained this indicates that there is cross polarity.

10.2.6 Interconnection of conductors

This type of fault is common in control systems and heating systems. The only way to check for it is to carry out a continuity test on the individual lengths of conductors while using the manufacturer's circuit diagram as guidance. For example, if the manufacturer's guidance tells you that terminal 12 at the control unit should be connected to terminal 4 at the pump, you will need to test the continuity between the conductors at these two terminals (with the cables removed from the terminals). If the reading shows good continuity, this indicates that the conductors have been in the correct location. You will need to continue testing the conductors until you find the one causing the problem.

Step 4: locating the fault

Once the type of fault has been identified, it must be accurately located.

One common method used to identify the location of a fault is called 'sectionalization' or the 'half split method'.

This method divides the circuit into smaller sections, which makes it easier to identify the exact location of the fault. The diagrams below show the simple steps required to locate a fault on a ring final circuit. This method can also be used on radial circuits.

10.2.7 Finding a fault on a ring final circuit

Figure 10.6 shows a ring final circuit with no fault indicated. It has been identified that there is a fault on this circuit, somewhere; using sectionalization, the fault can be located with accuracy.

▼ **Figure 10.6** Ring final circuit

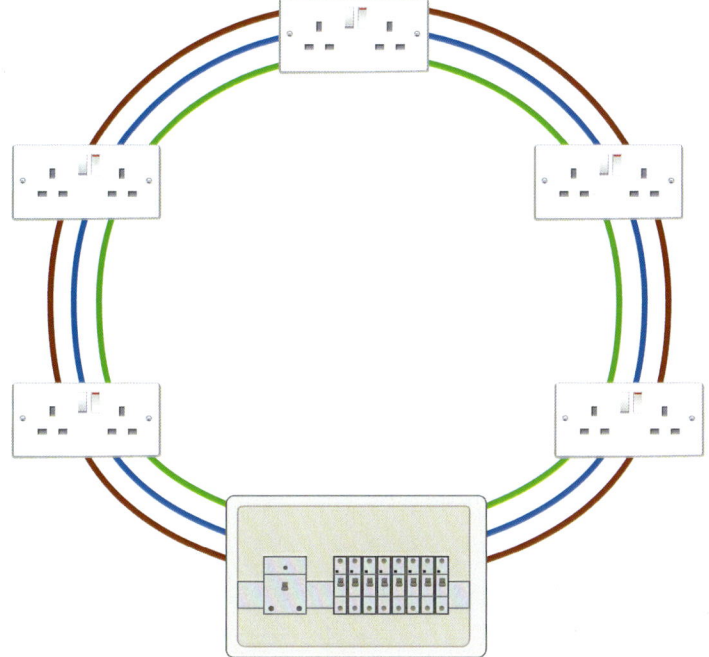

The following method should be used.

Carry out a risk assessment for the work to be done. Prior to beginning any test the power supply must be isolated, which requires the safe isolation procedure to be carried out. When the circuit has been successfully de-energized, the testing procedure can begin:

- **(a)** remove the conductors from terminals at the consumer unit. Once this has been done, you can carry out a test on the conductors and determine what type of fault there is on the circuit.
- **(b)** the first test is to check that the live conductors and cpcs are electrically separated (as they should be on a normal working circuit). This test will require the use of a continuity test instrument. Carry out the following test procedure:
  - **(i)** hold the probes of the tester onto the line and neutral conductors (L-N);
  - **(ii)** then hold the probes onto the line and cpc (L-E); and
  - **(iii)** then hold the probes onto the neutral and cpc (N-E).

10

On a ring final circuit, both legs of the ring should be tested.

On a circuit with no faults, the reading on a low ohms test will be high, for example, >1,999 Ω.

▼ **Figure 10.7** Ring final circuit, short circuit between line and neutral

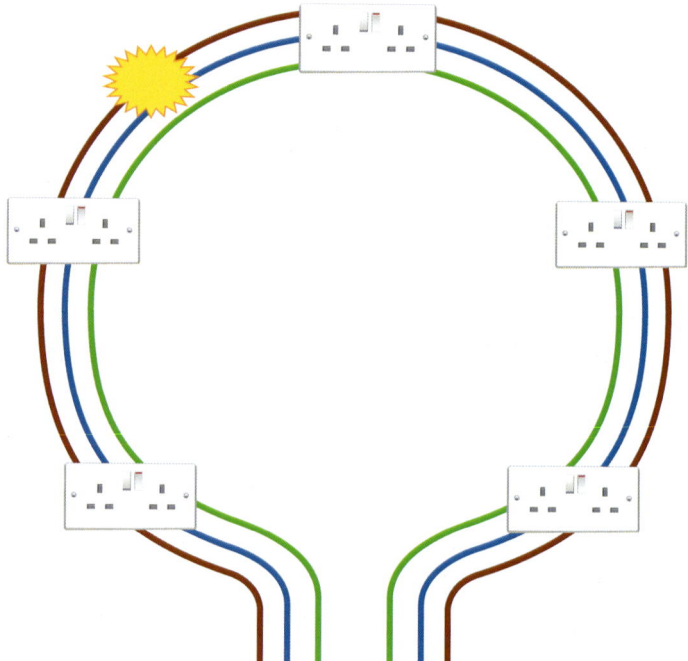

In this example, there is a fault between the line and the neutral conductors. When a continuity test is carried out, it will indicate a low resistance between these two conductors, indicating a short circuit between the line and neutral.

The next step is to find out where the fault is. To begin locating the fault, the circuit must be split into two parts (the half split method).

As you can see, using this method instantly splits the circuit into two parts, each half the size of the original – only this time, one half of the circuit has no fault.

By carrying out the same test as before, you can identify which half that is. The part of the circuit without a fault no longer needs testing and you have thus essentially reduced the area that needs testing in order to locate the fault by half.

▼ **Figure 10.8** Ring final circuit split into two parts

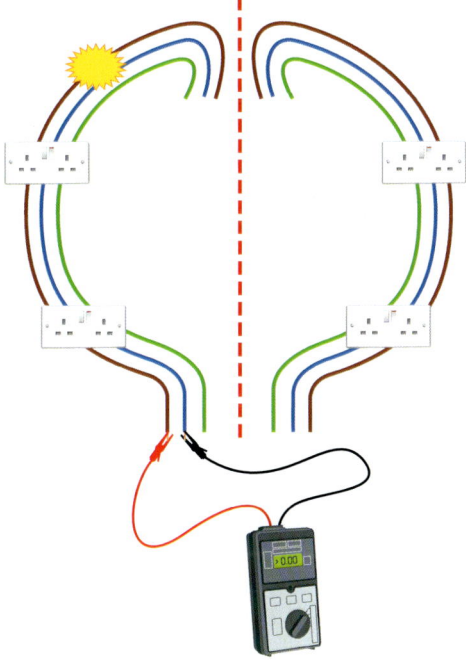

If you need to reduce the size of the circuit further, simply repeat the steps. You can continue splitting the circuit into smaller sections, re-testing each part until you have located the fault.

▼ **Figure 10.9** Ring final circuit split again

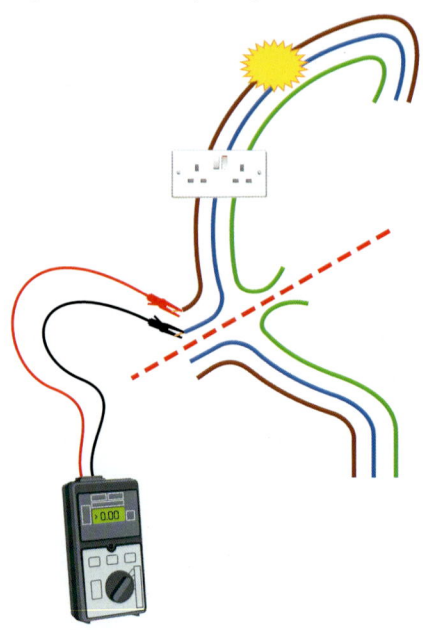

It may be that the fault is on a length of cable between two socket-outlets. This method will at least reduce the area in which you are fault-hunting.

As mentioned earlier in this section, this method can also be used for radial circuits. The principle remains the same: break the circuit down into smaller sections, so that the fault can be located.

There are various other ways to find the location of a fault. In most circumstances, it will involve segregating the circuit into smaller parts and testing the individual parts until the fault is located. It is not unusual for an electrician to install new cable to bypass part of a circuit that has a fault, as it may be difficult to get to the area of the fault, for example, under a tiled or fixed wooden floor. However, it is the responsibility of the electrician to ensure the safety of an installation before leaving it in the hands of ordinary persons.

Once the fault has been rectified, the entire part of the installation that was affected by the fault must be tested again, using the steps described in Section 9.

10

	Test your knowledge
1.	If a fault cannot be rectified immediately, how must the circuit be left?
2.	When testing EFLI, what would indicate that the installation may be satisfactory: a very high value or a very low value?
3.	When fault finding, you discover that there is a short circuit on a radial final circuit that supplies six BS 1363 socket-outlets. What is the first logical step you would take to begin locating the fault?
4.	What must be done once a fault has been rectified?

Common Calculations 11

This Section provides information on the following topics:
- ▶ Simple transposition
- ▶ Triangle method
- ▶ Resistors in series
- ▶ Resistors in parallel

11.1 Simple transposition

11.1.1 What is transposition?

When you need to calculate a value using an equation, you may need to transpose the formula in order to find the unknown. Calculations have rules: if you do something with one value, it will affect another. For example, if you increase the resistance of a circuit, you will reduce the amount of current (I amps) that can flow through that circuit. By moving the values around in the calculation, the answer can be found. The following rules must be applied, so that transposition can be used correctly:

  (a) the unknown should always be on its own, on the top line and on the left-hand side of the equation;
  (b) to get the unknown on its own, some of the values must be moved; and
  (c) when moving a value across the equals sign it must move either from above to below the line or from below to above the line.

For example:

3 × 6 = 9x

The first step is to move the unknown to the left-hand side of the equation:

9x = 3 × 6

Then draw an imaginary line underneath the right-hand side of the equation. Remember that the unknown is always on the top line:

9x = 3 × 6

To transpose this, you have to leave the unknown value on its own, so the 9 will be the logical value to move:

$$9x = \frac{3 \times 6}{9}$$

When the 9 is moved to the opposite side of the equals sign, it must go below the line, changing the equation to this:

$$x = \frac{3 \times 6}{9}$$

Once you have done the maths, you can find the answer:

$$x = \frac{18}{9}$$

$$x = 2$$

To check the answer, you can re-write the equation to look like this:

3 × 6 = 9 × 2

3 × 6 = 18 and 9 × 2 = 18

Remember that one side of an equation will always be equal to the other side!

11.2 Triangle method: voltage, current, resistance and power

The following triangles will help when calculating simple values for volts, amps, ohms and watts.

They are the first of many formulae you will learn as you progress in your electrotechnical career.

V for voltage in volts (V)	**I** for current in amps (A)
R for resistance in ohms (Ω)	**P** for power in watts (W)

You can use the triangles below by covering the value you want to find. For example, if you already know the value of V and R and you want to find I, you simply cover I on the triangle. This will indicate that you need to divide V by R, which will give you the value for I. The same principle applies when trying to find any other values.

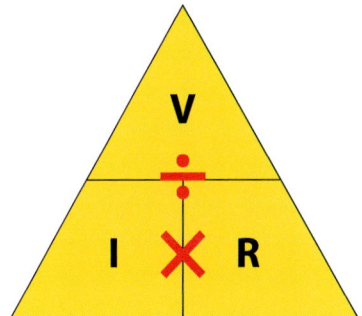 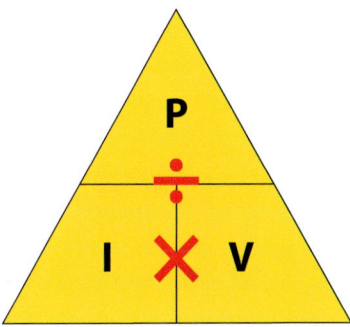

Ohm's Law (VIR)	Power Law (PIV)
In this triangle you can see that:	In this triangle you can see that:
R = V ÷ I	V = P ÷ I
I = V ÷ R	I = P ÷ V
V = I × R	P = I × V

11.3 Resistors

NOTE: It's important to use the correct terminology when citing voltage and current. Current is always 'through' a component and voltage is always 'across' a component.

11.3.1 Resistors in series

In a series circuit, the current remains constant through each resistor. If, for example, there is 2 A through R_1, there will be 2 A through R_2 and 2 A through R_3.

When adding up the total resistance of a series circuit, you simply add all the values together.

To calculate the total resistance of the following circuit, add all the resistor values together:

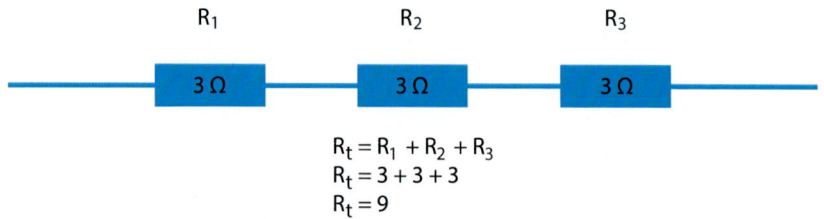

$R_t = R_1 + R_2 + R_3$
$R_t = 3 + 3 + 3$
$R_t = 9$

The total resistance (R_t; t for 'total') is therefore 9 Ω.

11.3.2 Resistors in parallel

In a parallel circuit, the voltage remains constant across each resistor. If, for example, there is a 20 V supply, there will be 20 V across R_1 and 20 V across R_2.

When resistors are connected in parallel, an extra path is being created for the current to flow. So, instead of increasing the overall resistance, as happens in a series circuit, the overall resistance is reduced.

Imagine that the resistors are narrow alleyways and the current is a group of people trying to get through. If there is only one alleyway, it will take twice as long for a group of people to walk through than it would if there were two alleyways.

To calculate the total resistance of a resistor network connected in parallel, you need to establish the current through each resistor. The voltage remains constant across each resistor in a parallel circuit.

Remember Ohm's Law: $I = \dfrac{V}{R}$

Which looks like: $I = \dfrac{20}{5}$

$I = 4$

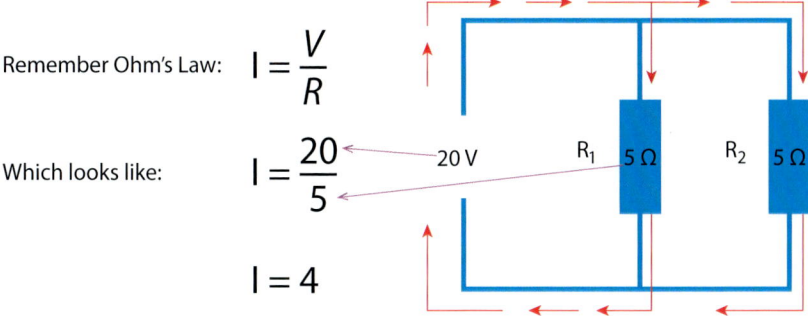

The current flow through each resistor is thus 4 A.

To calculate the total current for the circuit, the current flow through each resistor must be added together. In the example above, it would be:

4 + 4 = 8 A

A total of 8 A is thus flowing through the circuit. Now that we have the total current for the circuit, we can work out the total resistance of the circuit, using Ohm's law:

$\dfrac{20}{8} = 2.5 \ \Omega$

You can see that the total resistance is **reduced** when an additional resistor is connected in parallel in the circuit.

What if we don't know the value of V?

There is another way of working out the total resistance of resistors in parallel. If only the resistance values are known, you can use the following equation:

$$\frac{1}{R_t} = \frac{1}{R_1} + \frac{1}{R_2}$$

$$\frac{1}{R_t} = \frac{1}{5} + \frac{1}{5}$$

$$\frac{1}{R_t} = \frac{1+1}{5}$$

$$\frac{1}{R_t} = \frac{2}{5} = 0.4$$

Remember that the value you are trying to find is not a fraction, so you need to carry out one more calculation:

If

$$\frac{2}{5}$$

is equal to

$$\frac{1}{R_t}$$

then

$$\frac{R_t}{1}$$

is equal to

$$\frac{5}{2}$$

so this essentially means that

$$R_t = \frac{5}{2}$$

$$R_t = 2.5 \, \Omega$$

When using a calculator, you will need to use the X⁻¹ button and the S↔D button.

The sequence will look like this:

5 → x⁻¹ → + → 5 → x⁻¹ → = → 2/5 → SD → 0.4 → x⁻¹ → = → 2.5

In this example, the answer is:

2.5 Ω

This answer should always be a lower resistance than the smallest resistor on the circuit.

Finding a percentage (%) the easy way

Let's say you want to determine what 4 % V drop is on a circuit supplied at 230 V. The easiest way to find out looks like this:

$$VD = \frac{V \times \%}{100}$$

$$\frac{230 \times 4}{100} = 9.2$$

When using the % function on a calculator, you should enter the values like this:

230 × 4 % = 9.2

11

Test your knowledge

1.	Try transposing these to find the value of x:

4 × 6 = 2x	9 × 3 = 3x	4 × 2 × 3 = 2 × 4x

2.	What is power measured in?
3.	What formula do you use to find V if you have the values of R and I?
4.	A 230 V electrical appliance is rated at 460 W. What will the current flow be in amps when the electrical supply is 230 V?
5.	Calculate the total resistance (R_t) of these circuits:

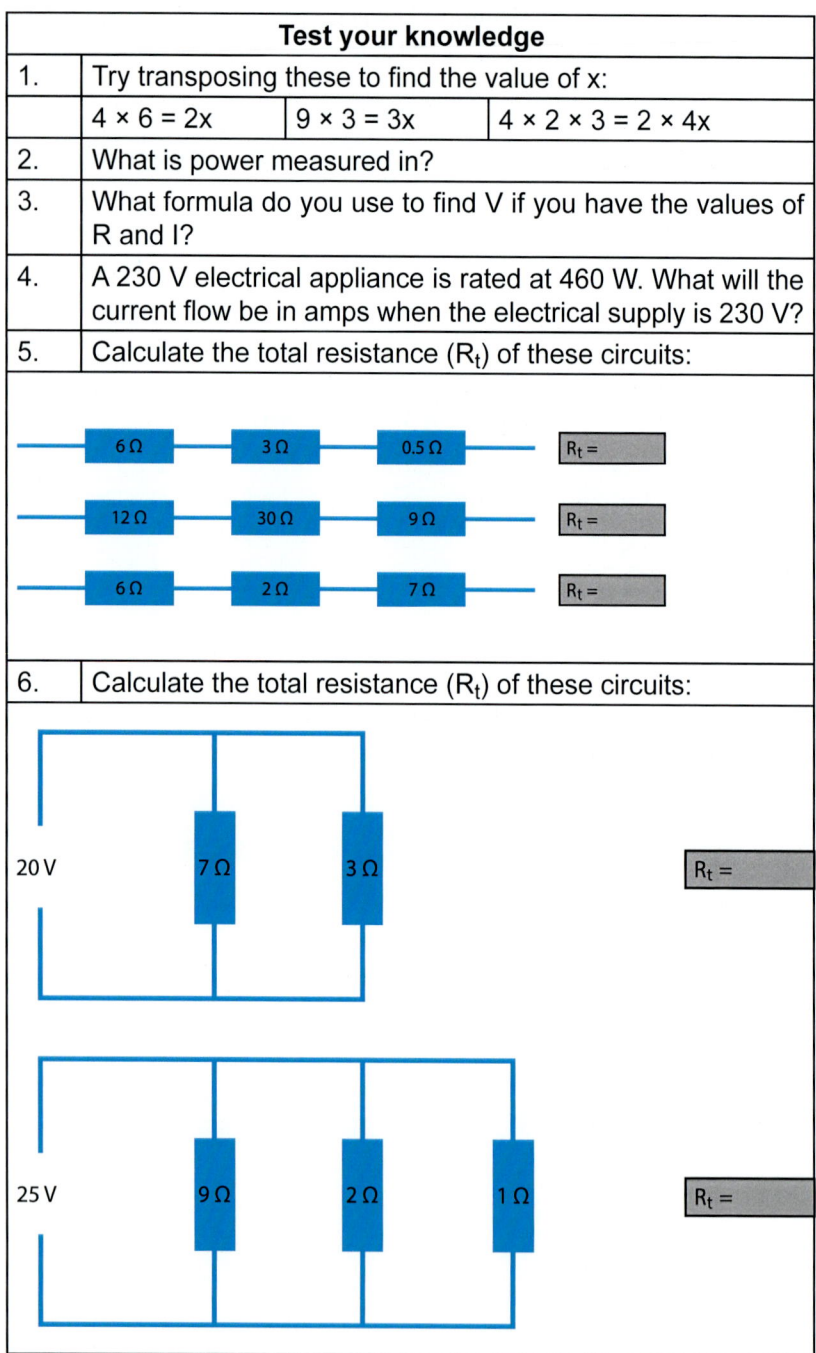

6. Calculate the total resistance (R_t) of these circuits:

Diversity   12

This Section provides information on the following topics:
- ▶ What diversity is and why we need it
- ▶ Calculating diversity
- ▶ Applying diversity to various installations

12.1 What is diversity?

Diversity is the term used to describe the calculation that determines the likely maximum demand at any one point on an electrical circuit or installation (not to be confused with the maximum connected load). This enables electricians to design an electrical installation on the assumption that every piece of electrical equipment will not be switched on and running at full capacity all the time. Cables can withstand high currents for short periods of time and in some cases, for example, when designing a circuit for a motor or for equipment that uses a large amount of power for a short period of time, a smaller cross-sectional area (csa) can be used.

If every installation was designed to carry the maximum connected load, the cost to customers would be extremely high and wiring systems would be over engineered. In some ways, this is similar to the way in which a car is designed and manufactured: the vehicle is not intended to be driven at maximum speed everywhere it goes. Similarly, it would be unrealistic to assume that all electrical equipment will be switched on at full power all the time.

12.2 Calculating diversity

Because diversity is worked out differently depending on the characteristics of the circuit, the easiest way to explain the calculation is through an example. When designing an installation to supply a cooker hob and oven, the following method can be used.

The oven is rated at 4,000 W and you have four rings on the hob:

(a) 800 W;
(b) 1,000 W × 2; and
(c) 2,000 W.

Looking at these values, the assumed design current would be:

800 + (1,000 × 2) + 2,000 + 4,000

Total connected load = 8,800 W (8.8 kW)

We know that watts is the measurement of power, so, by using the power triangle from Section 11, the calculation to determine the design current

$$I_b = \frac{P}{V}$$

would look like this:

$$I_b = \frac{8800}{230}$$

so:

I_b = 38.26 A

However, this is the total connected load of the hob and oven when every element is turned on. The likelihood of this happening for a long period of time is extremely small; if the electrician were to install a cable to meet these requirements, the cable would be oversized and would only ever reach a fraction of its current-carrying capacity, as well as costing the customer a lot more than necessary. If a protective device is chosen based on these requirements, it may not be suitable for its intended purpose.

Using Tables A1 and A2 from the *On-Site Guide*, it is possible to determine roughly how much actual current the installation will be drawing. This is known as 'diversified load'.

▼ **Figure 12.1** Extract from Table A1 of the *On-Site Guide*

Point of utilization or current-using equipment	Current demand to be assumed
Socket-outlets other than 2 A socket-outlets and other than 13 A socket-outlets See NOTE 1	Rated current
2 A socket-outlets	At least 0.5 A
Lighting outlet See NOTE 2	Current equivalent to the connected load, with a minimum of 100 W per lampholder
Electric clock, shaver supply unit (complying with BS EN 61558-2-5), shaver socket-outlet (complying with BS 4573), bell transformer, and current-using equipment of a rating not greater than 5 VA	May be neglected for the purpose of this assessment
Household cooking appliance	The first 10 A of the rated current plus 30 % of the remainder of the rated current plus 5 A if a socket-outlet is incorporated in the control unit
All other stationary equipment	British Standard rated current, or normal current

Using the table, we can begin to calculate the maximum demand (connected load including an allowance made for diversity).

For the calculation of the hob and oven:

The design current is 38.26 A (from our earlier calculation), with the first 10 A taken at 100 % of its value. This means that you take 10 A from the value and put it to the side, ready to add to the remaining value once the following calculations have been carried out.

So:

38.26 − 10 = 28.26 A

The remaining 28.26 A is taken at 30 %. The easiest way to work out 30 % of 28.26 is:

(28.26 ÷ 100) × 30 (or, if you have a calculator, 28.26 × 30 %) = 8.47.

Once this value has been calculated, it can be added to the 10 A that was put on the side at the beginning: 8.47 + 10 = 18.47 A.

▼ **Figure 12.2** Process sequence for calculating diversity for a cooker circuit

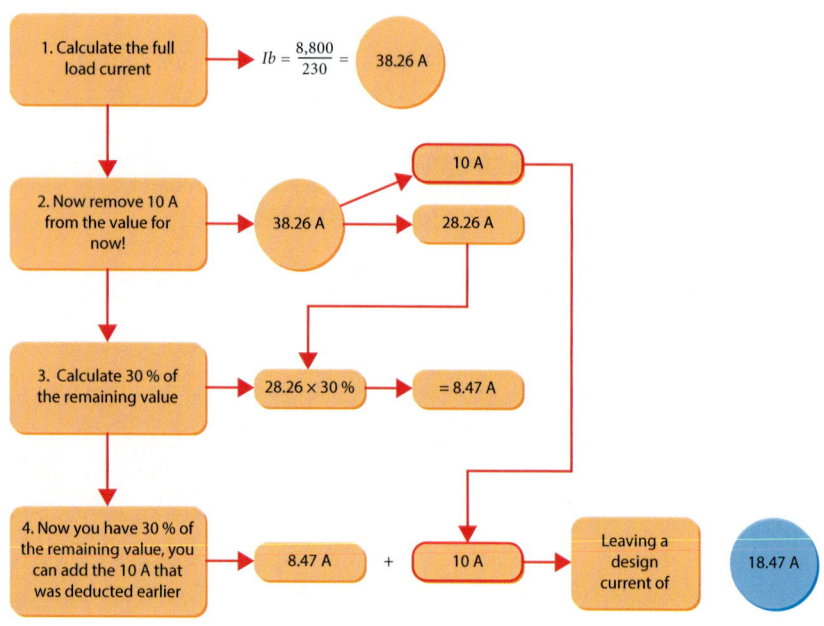

What next?

The value remaining after the calculation has been completed is used to select the rating of the protective device. In this case, the value is 18.47 A, so a protective device of this rating or the next available higher rating must be selected, such as a 20 A circuit-breaker (see the relevant Standard, BS EN 60898).

12.3 Diversity for ring final circuits

Each socket-outlet is designed for a plug of up to 13 A: for a double socket, that is potentially 26 A. In the circuit shown below, the total current that may be drawn from this circuit is 130 A.

▼ **Figure 12.3** Ring final circuit

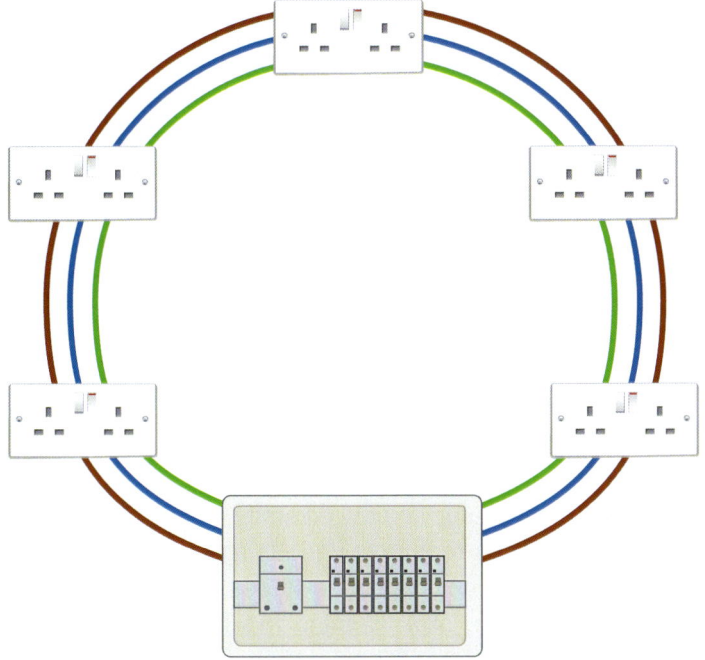

It would be unrealistic to install cable and circuit-breakers with a rating of 130 A for a ring final circuit, as it is extremely unlikely that it would ever be used to this extent. Therefore, diversity can be taken into account. Table H2.1 from the *On-Site Guide* provides guidance on the maximum floor area that can be served with BS 1363 socket-outlets. Diversity has already been taken into account in this table and does not need to be applied again.

▼ **Figure 12.4** Table H2.1 from the *On-Site Guide*

Type of Circuit		Overcurrent protective device rating (A)	Minimum live conductor cross-sectional area* (mm²)		Maximum floor area served (m²)
			Copper conductor thermoplastic or thermosetting insulated cables	Copper conductor mineral insulated cables	
1	2	3	4	5	6
A1	Ring	30 or 32	2.5	1.5	100
A2	Radial	30 or 32	4	2.5	75
A3	Radial	20	2.5	1.5	50

12.3.1 Kitchen appliance demands

The image below shows how diversity can be considered for circuits that are designed to supply various appliances that have a varying load profile depending on the stages of their cycle. For example, the washing cycle of a washing machine begins with it turning to take on water and wet the garments within. It then heats the water, has various washing and rinsing cycles, and ends with a long spin cycle to remove the water. The appliance, therefore, will not run at maximum power for the entire duration of the wash.

You will notice on the image that the total load increases to above 30 A on two occasions for a short period. Most protective devices will allow this small overload current for a limited time, and it is considered an acceptable practice when designing an installation. In the circuit below, a 32 A circuit-breaker (BS EN 60898) can therefore be considered for a protective device, as the load is only expected to be higher than 32 A for a very short time.

▼ **Figure 12.5** Diversity used in different circuits

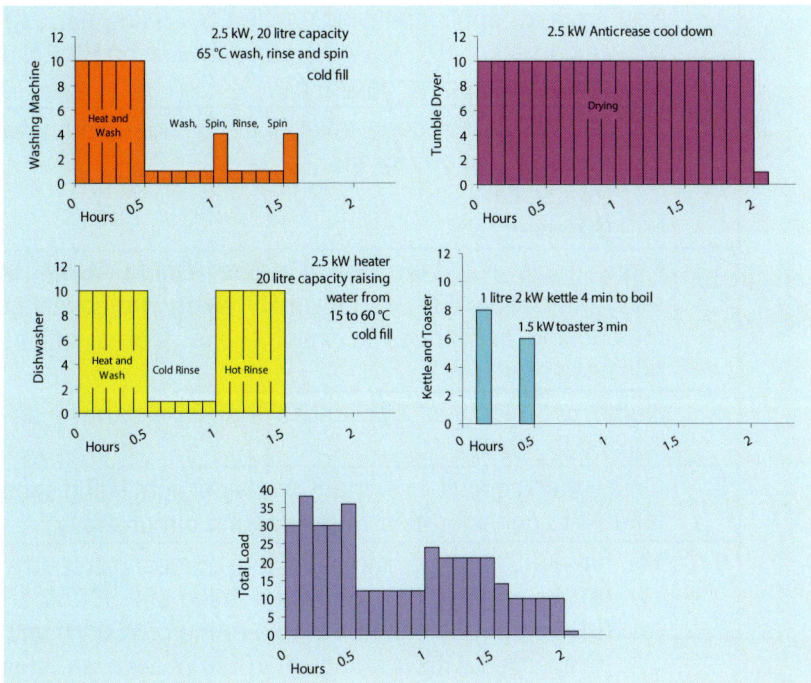

12.3.2 When can diversity not be applied?

When designing an installation for equipment such as a single water heater (thermostatically controlled) or an electric shower, the design current will be constant and diversity should not be applied. This type of circuit will be either on at full load or completely off-load. This means that, when a water heater is switched on, it will use the same amount of power until it is switched off again. The demand will not vary as it does with cooker circuits and washing machines, etc.

The important thing to remember about diversity is that it is specific to every installation and the electrician/designer must consider what the likely demand is going to be on the circuit when selecting the appropriate cable and protective devices.

12

Test your knowledge

1.	You have an appliance which, when operating normally, will draw a current of 15.5 A. What rating should the BS EN 60898 overcurrent protective device be?
2.	If diversity has not been taken into consideration, what are the main implications for the: **(a)** installer? **(b)** customer?
3.	From Table A2 from the *On-Site Guide*, what is the recommended value, with diversity taken into consideration, for cooking appliances in small shops, stores, offices and business premises?
4.	Name two electrical appliances that cannot have diversity applied to them.
5.	On a lighting circuit, how many watts per light fitting should be taken into consideration when working out diversity?
6.	You have a 230 V hob and oven, in a domestic property, with: **(a)** two rings on the hob with a rating of 2,000 W; **(b)** two rings on the hob with a rating of 750 W; and **(c)** an oven with a rating of 3,000 W, with an additional socket-outlet. What size circuit-breaker should be installed? Show all working.

Prosumer's Electrical Installations   13

13.1 Prosumer's installations

Chapter 82 of BS 7671 provides the additional requirements, measures and recommendations for the design, erection and verification of all types of low voltage (LV) electrical installations identified in the Scope (Chapter 11) of BS 7671, including the local production and/or storage of energy in order to ensure compatibility with existing and future ways to deliver electrical energy to:

(a) current-using equipment; or

(b) the public network by means of local sources of electrical energy.

Such electrical installations are designated as prosumer's electrical installations (PEIs).

A PEI is a set of electrical equipment having the following functions (see Figure 13.1):

(a) supply (e.g. connection to a public power supply, local generator, photovoltaic (PV) systems, wind turbines, batteries);

(b) distribution (e.g. distribution panel, wiring systems);

(c) consumption (e.g. motors, heating systems, lighting, lifts); and

(d) electrical energy management system (EEMS) (e.g. load shedding equipment, monitoring device, etc.).

NOTE: A battery can be considered as a generator and as a load.

13

An uninterruptible power supply (UPS) is not considered to form part of a PEI as the purpose of a UPS is to supply only downstream critical loads and it is not capable of reverse feeding the public network and/or current-using equipment in the upstream part of the electrical installation.

▼ **Figure 13.1** Example of prosumer's electrical installations

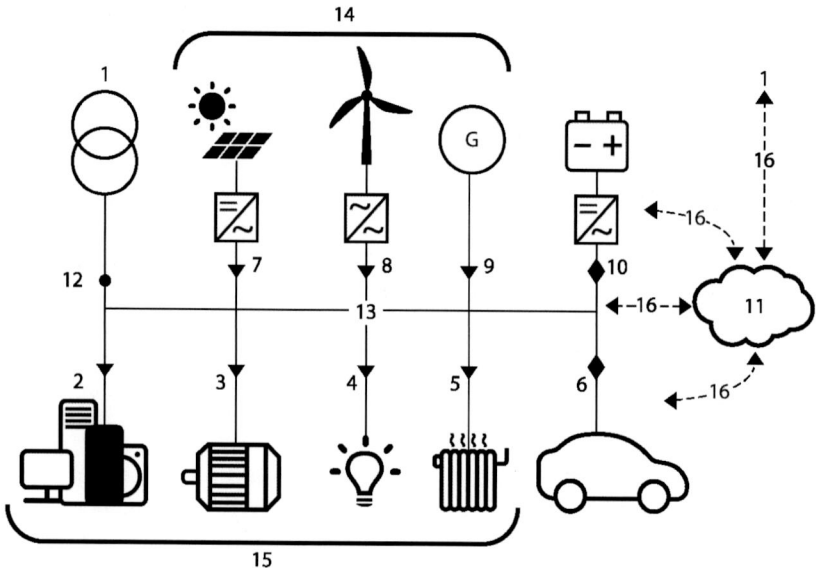

Key

1	Public network	9	Other generators
2	Home appliances and electronic devices	10	Electric storage
3	Motors	11	EEMS
4	Lighting	12	Origin of installation
5	Heaters	13	Local distribution
6	Electric vehicles	14	Local generation
7	Solar inverter	15	Local consumption
8	Wind inverter	16	Management signals

In a PEI, an installation owner may consider independently the supervision and the control of different power supplies connected to the electrical installation in order to supply efficiently and cost-effectively all the electrical loads connected to this LV electrical installation.

Chapter 82 of BS 7671 provides requirements for PEIs to achieve safe operation, sustainability and efficient use of energy when integrated into smart grids.

It does not cover electrical sources for safety services, including associated electrical installations and standby electrical supply systems for a secure continuity of supply, which are operated only occasionally and for short periods in parallel with the distribution grid for testing purposes.

13.2 Types of prosumer's electrical installations

There are different types of PEI:

> **(a)** individual (see Regulation 824.3);
> **(b)** collective (see Regulation 824.4); and
> **(c)** shared (see Regulation 824.5).

13.3 Operating modes

The main operating modes that may be adopted for each type of PEI (individual, collective or shared) are:

> **(a)** direct feeding mode;
> **(b)** reverse feeding mode; and
> **(c)** island mode.

Storage units can:

> **(a)** supply current-using equipment; or
> **(b)** be charged by local power supplies; or
> **(c)** be charged by the public network;

except when operating in island mode.

Local power supplies can supply:

(a) current-using equipment; or
(b) local storage units; or
(c) the public network;

except when operating in island mode.

Transfer to and from the direct feeding mode to island mode and vice versa can be achieved by operating the switching device for islanding; this can be either directly controlled (manually or remotely) or automatically controlled.

Switching from one mode to another can be done if the generators and/or converters are synchronized with the network (providing the requirements of Regulation 551 are met).

Selection of the possible operating modes may depend on the contract with the DNO (distribution network operator).

Technical requirements for the design of the PEI according to the selected operating mode are provided in Regulation 826 of BS 7671. In addition:

(a) the general principles and types of PEI are described in BS 7671: Annex A82; and
(b) examples of operating modes are contained in BS 7671: Annex B82.

	Test your knowledge
1.	Where in BS 7671 will you find the information and requirements for prosumer's installations?
2.	A PEI is a set of electrical equipment performing what four functions?
3.	Which two functions can a battery perform?
4.	What are the three operating modes of a PEI?
5.	What functions can storage units perform?
6.	What functions can the local power supplies perform?
7.	In which operating mode can the PEI not perform the functions listed in questions 5 & 6?
8.	What does the term EEMS stand for?

Special installations or locations   A

As mentioned at the beginning of this Guide, the IET Wiring Regulations have been around since 1882. It goes without saying that a lot has changed since then. Electricity has taken over almost every possible location there is, including solar photovoltaic (PV) systems on roof tops; swimming pools; electric vehicle (EV) charging points on the street or in domestic dwellings, etc.

Even 50 years ago it was impossible to imagine the extent to which electricity would be used today. Part 7 of the IET Wiring Regulations takes into consideration some of the special installations or locations where electrical systems will be installed. It is not the intention to create completely new regulations to suit various different locations, but rather to identify which existing regulations need to be considered and, if necessary, modified, when installing an electrical system in a special location. For one reason or another, all of the special installations or locations mentioned in Part 7 of the IET Wiring Regulations need additional consideration when electrical equipment and wiring systems are being installed. It needs to be borne in mind that Part 7 should be used in conjunction with the other parts of the IET Wiring Regulations and not as the sole requirements for that installation.

In this Appendix, we only briefly introduce you to the areas that qualify as special locations for installations. The IET's Guidance Note 7 provides further information and guidance on all the special locations and installations as set out in the IET Wiring Regulations.

A

It is recommended that you consult Guidance Note 7 and the IET Wiring Regulations should you find yourself working on any of the areas listed in the table below.

There are currently 20 special installations or locations covered in Part 7

Location	Section	Description
Locations containing a bath or shower	701	What needs to be borne in mind is that we conduct electricity more easily when we come into contact with a live part when we are wet. The following factors should be considered when installing electrical systems in a location containing a bath or shower: (a) the likelihood that persons will not be wearing any clothing or footwear – this increases the contact area they have with surfaces at earth potential; (b) the presence of water, which can reduce the contact resistance between parts that may become live (under fault conditions) and the human body; and (c) the fact that when people are immersed in water, their total body resistance to current is reduced – more information on the resistance of the body can be found in IEC/TS 60479. Figures 701.1 of the IET Wiring Regulations and 8.1 (i) and (ii) of the *On-Site Guide* show the zone dimensions in a location containing a bath or shower.
Swimming pools and other basins	702	As mentioned in 701 above, the resistance between the body and earth is reduced when a person is unclothed or partially unclothed, especially when not wearing some form of footwear and when the body is likely to be wet. 702 applies to the basins of swimming pools, fountains and paddling pools. It also covers the surrounding area of these basins that have allocated zones.
Rooms and cabins containing sauna heaters	703	If there are showers or similar facilities installed, the requirements of 701 also apply. Temperature and extreme humidity are factors that really need to be considered in locations containing sauna heaters. 703 provides the installation requirements and the zone dimensions.

Location	Section	Description
Construction and demolition site installations	704	Construction sites are potentially dangerous in many ways. Four factors contribute to the high risk of electric shock on a construction site: (a) the possibility of damage to cables and equipment; (b) the widespread use of hand tools with trailing leads (this problem is mitigated by the increasing use of battery-operated tools); (c) the accessibility of many extraneous-conductive-parts, which cannot practically be bonded; and (d) the fact that the works are generally open to the elements. Measures to reduce the risks include: (a) the preference, as stated in the IET Wiring Regulations, for the reduced low voltage system (110 V AC, which, at single phase, is 55 V to earthed midpoint of the transformer) to supply portable handlamps for general use and portable hand tools and local lighting up to 2 kW, while separated extra-low voltage (SELV) is strongly preferred for portable handlamps in confined or damp locations; (b) the avoidance of the use of a PME (TN-C-S) earthing facility for the means of earthing for a construction site installation unless all extraneous-conductive-parts are reliably connected to the main earthing terminal (MET); (c) the protective measures of obstacles, placing out of reach, non-conducting location and prohibiting or not permitting earth-free local equipotential bonding; (d) requiring that equipment for external use should be at least IP44; and (e) the additional testing of residual current devices (RCDs).

Location	Section	Description
Agricultural and horticultural premises	705	705 provides the requirements for agricultural and horticultural premises, such as farmyards, cattle yards, slaughterhouses, greenhouses, etc. In circuits, whatever the type of earthing system, the following protective devices shall be provided: (a) circuits supplying socket-outlets up to 32 A: a 30 mA RCD; (b) circuits supplying socket-outlets over 32 A: a 100 mA RCD; and (c) all other circuits: a 300 mA RCD.

Conducting locations with restricted movement	706	The particular requirements of 706 apply to fixed equipment in conducting locations where movement of persons is restricted by the location, and supplies to mobile equipment in these locations. Such locations are where movement is heavily restricted and the area is metallic and electrically conductive. Examples of this type of location are working at high level on a steel walkway or gantry or within a stainless steel storage vessel.
		
		There are particular requirements for protection against electric shock. In conducting locations with restricted movement, the following protective measures apply to circuits supplying the following current-using equipment:
		
		(a) supply to a hand-held tool or an item of mobile equipment:
		  (i) electrical separation; or
		  (ii) SELV;
		(b) supply to handlamps: SELV; and
		(c) supply to fixed equipment:
		  (i) automatic disconnection with supplementary equipotential bonding; or
		  (ii) the use of Class II equipment and additional protection by the use of RCDs at 30 mA; or
		  (iii) electrical separation; or
		  (iv) SELV; or
		  (v) protective extra-low voltage (PELV).

Location	Section	Description
Marinas and similar locations	709	The environment of a marina or yachting harbour is harsh for electrical equipment. The water, salt, high winds and movement of structures accelerate deterioration of the installation. The presence of salt water, dissimilar metals and a potential for leakage currents increases the rate of corrosion. There are also increased electric shock risks associated with a wet environment, through reduction in body resistance and contact with earth potential. Particular requirements to reduce the risks include: **(a)** prohibition of a TN-C-S system for the supply to a boat; **(b)** additional protection by 30 mA RCDs; and **(c)** socket-outlets to be installed at not less than 1 m above the highest water level (this may be reduced if additional measures are taken).

A

Location	Section	Description
Medical locations	710	710 provides requirements for hospitals, private clinics, medical and dental practices, healthcare centres and dedicated medical rooms in the work place, and can also be used in veterinary clinics.

Risks in this location may arise from the following factors:

(a) the natural protection of the human body against electric shock is considerably reduced when certain clinical procedures are being performed on it.

(b) patients under treatment may have their skin resistance broken or their defensive capacity either reduced by medication or nullified while anaesthetized.

(c) in addition to the risk from electric shock, some equipment, such as life-support equipment or surgical equipment, performs such vital functions that loss of the electrical supply would pose an unacceptable risk to patients. Medical locations where such vital equipment is used require secure electrical supplies.

To protect patients from possible electrical hazards, 710 requires additional protective measures to be applied in medical locations.

These include:
(a) particular requirements for protection against electric shock;
(b) medical IT systems (see Guidance Note 7);
(c) requirements for supplementary equipotential bonding;
(d) additional requirements for the selection and erection of electrical equipment including switchgear and controlgear;
(e) safety services, including the sources, such as generators; and
(f) detailed requirements for safety lighting.

Location	Section	Description
Exhibitions, shows and stands	711	711 applies to temporary electrical installations in exhibitions, shows and stands. It does not apply to: **(a)** the building in which the event is occurring; **(b)** electrical systems used in the structures, sets, etc. of production and similar activities of the entertainment industry; or **(c)** the exhibits themselves. The risks associated with exhibitions, shows and stands are those of electric shock and fire. These arise from: **(a)** the temporary nature of the installation; **(b)** the lack of permanent structures; **(c)** severe mechanical stresses; and **(d)** access to the general public. Because of these increased risks, additional measures are required: **(a)** a cable intended to supply temporary structures shall be protected at its origin by an RCD with a maximum residual operating current of 300 mA; **(b)** all metallic structural parts that are accessible within the stand, etc. are to be bonded; **(c)** additional protection is required for all final circuits (other than for emergency lighting) and socket-outlets up to 32 A by an RCD, in accordance with Regulation 415.1.1; **(d)** TN-C-S shall not be used (there are exceptions); and **(e)** every temporary structure, such as a vehicle, stand or unit, intended to be occupied by one specific user, and each distribution circuit supplying outdoor installations, shall be provided with its own readily accessible and properly identifiable means of isolation.

Location	Section	Description
Solar power supply systems	712	712 applies to the electrical installations of PV power supply systems. Some key points: (a) PV modules generate electricity at DC. It is connected to an inverter that converts DC to AC to allow the connection to the AC electrical supply of the building. The DC side of the installation, i.e. the PV modules, is considered to be energized even when the system is disconnected from the AC side. Remember that when it's daylight, the modules are generating. (b) to allow for maintenance, a means of isolating the AC and DC sides of the PV convertor shall be provided. (c) there are also many other additional requirements (see Section 712 in the IET Wiring Regulations).
Outdoor lighting installations	714	Outdoor lighting installations include those for roads, parks, car parks, gardens, places open to the public, sporting areas, illumination of monuments, floodlighting. Road signs are also included, but not road traffic signal systems. In addition, other lighting arrangements in places such as telephone kiosks, bus shelters, advertising panels and town plans are covered and all require RCD protection. The following are excluded: (a) temporary festoon lighting; (b) luminaires fixed to the outside of a building and supplied directly from the internal wiring of that building; and (c) road traffic signal systems. Section 714 does not include temporary festoon lighting, chiefly because it will be erected on an ad-hoc basis and will not form part of the fixed installation. Furthermore, luminaires fixed to the outside of a building and supplied directly from the internal wiring of that building are not included, as these are considered part of the building's electrical installation.

Location	Section	Description
ELV lighting installation	715	The particular requirements of these regulations apply to ELV installations supplied from sources with a maximum rated voltage of 50 V AC rms or 120 V DC. The regulations include requirements for: **(a)** protection against electric shock (SELV); **(b)** protection against the risk of fire due to short circuit; **(c)** types of wiring systems, including special requirements where bare conductors are used; **(d)** types of transformers and converters; and **(e)** requirements for suspended systems.

A

Location	Section	Description
Mobile or transportable units	717	The particular requirements of 717 apply to mobile or transportable units. These may be self-propelled, towed or transportable containers or cabins. Examples of the units include technical and facilities vehicles for the entertainment industry, medical services units, advertising units, fire-fighting units, workshops, offices, transportable catering units, etc.

The risks associated with mobile and transportable units include:

  (a) loss of connection to earth due to the use of temporary supply cables;
  (b) risks arising from the connection to different national and local electricity distribution networks;
  (c) open-circuit faults of the protective earth and neutral (PEN) conductor of PME (TN-C-S) supplies raising the potential of all metalwork (including that of the unit) to dangerous levels;
  (d) risk of shock arising from high functional currents flowing in protective conductors; and
  (e) vibration while the vehicle or trailer is in motion, or while a transportable unit is being moved, causing faults within the unit installation.

Some of the requirements to reduce these risks include:

  (a) additional protection by an RCD;
  (b) accessible conductive parts of the unit to be connected through the main equipotential bonding to the MET within the unit; and
  (c) identification adjacent to the supply inlet for:
   (i) the type of supply that may be connected;
   (ii) the voltage rating of the unit;
   (iii) the number of phases and their configuration;
   (iv) the onboard earthing arrangement; and
   (v) the maximum power required by the unit.

A

Location	Section	Description
Electrical installations in caravan/camping parks, caravans and motor caravans	708 and 721	The particular requirements of 708 apply to the electrical installations in caravan/camping parks and similar locations providing connection points for supplying leisure accommodation vehicles (including caravans) and tents. The particular requirements of 721 apply to the electrical installations of caravans and motor caravans at nominal voltages not exceeding 230/440 V AC or 48 V DC. Note that there are some exclusions. The risks specifically associated with installations in caravan parks, caravans and motor caravans arise from: (a) open-circuit faults of the PEN conductor of PME (TN-C-S) supplies raising the potential to true Earth of all metalwork to dangerous levels; (b) incorrect polarity at the pitch supply point; (c) possible loss of earthing due to long supply cable runs, connecting devices exposed to weather and flexible cable connections liable to mechanical damage; and (d) vibration while the vehicle is moving, causing faults within the caravan installation. Particular requirements to reduce the above risks include: (a) prohibition of the connection of exposed- and extraneous-conductive-parts of a caravan or motor caravan to a PME (TN-C-S) terminal. Where the supply to the site is PME (TN-C-S), the earthing arrangement at the pitch supply point shall form part of a TT system. (b) additional protection by 30 mA RCDs in both the vehicle and the pitch supply point. (c) double-pole isolating switch and circuit-breakers protecting the final circuit in the tent, caravan or motor caravan. (d) internal wiring of the tent, caravan or motor caravan by flexible or stranded cables of a cross-sectional area (csa) of 1.5 mm² or greater, plus additional cable supports and the segregation of low-voltage (LV) and ELV circuits.

Location	Section	Description
Electric vehicle charging installations	722	722 specifies the requirements for charging supplies to electric vehicles. It includes requirements for the type and current rating of socket-outlets, RCD protection, measures of protection against electric shock, earthing requirements, IP rating of equipment, impact protection against mechanical damage, isolation and switching and fixing arrangements, etc.
		An important point is that Regulation 722.411.4.1 does not allow PME (TN-C-S) as a means of earthing for an electric vehicle charging point where the charging point or the vehicle is located outdoors, except where dangerous touch voltages are not possible or will be disconnected before a dangerous voltage occurs, or a supplementary earth electrode is installed. Protection against electric shock may also be provided by the use of an alternative device which does not result in a lesser degree of safety. The equivalent means of functionality could be included within the charging equipment.
		Another key point is that socket-outlets must be fit for purpose. They must be suitable for the load and for the external influences such as protection against mechanical damage and ingress of water. 722 requires a degree of protection of at least IP44 (protected against solid objects over 1 mm and protected against water splashed from all directions) where the equipment is installed outdoors. The use of 13 A extension leads, for example, would not meet the requirements of 722; however, the use of tethered vehicle connectors is allowed.

Location	Section	Description
Operating and maintenance gangways	729	729 applies to basic protection and other aspects in restricted access areas, for example, switchrooms with switchgear and controlgear assemblies with a need for operating or maintenance gangways for authorized persons. 729 contains requirements for: **(a)** restricted access areas where basic protection is provided by barriers or enclosures; and **(b)** restricted access areas where the protective measure of obstacles is applied. 729 gives examples of positioning doors in long closed restricted access areas.
Onshore units of electrical shore connections for inland navigation vessels	730	Section 730 applies to onshore installations that are dedicated to the supply of inland navigation vessels for commercial and administrative purposes, berthed in ports and berths. Most, if not all of the measures used to reduce the risks in marinas apply equally to electrical shore connections for inland navigation vessels. One of the major differences between supplies to vessels in a typical marina and electrical shore connections for inland navigation vessels is the size of the supply needed. For example, vessels used on inland waterways in Europe can weigh up to 10,000 tonnes – considerably larger than the average size of vessel used in a marina, which are generally small recreational craft (up to 24 m long).
Temporary electrical installations for structures, amusement devices and booths at fairgrounds, amusement parks and circuses	740	740 specifies the minimum electrical installation requirements to facilitate the safe design, installation and operation of temporary erected mobile or transportable electrical machines and structures that incorporate electrical equipment. The machines and structures are intended to be installed repeatedly, without loss of safety, in a temporary capacity at fairgrounds, amusement parks, circuses or similar places. The permanent electrical installation is excluded from the scope. 740 does not apply to the internal electrical wiring of the machines.

Location	Section	Description
Heating cables and embedded heating systems	753	753 applies to the installation of embedded electric heating systems for surface heating. It also applies to electric heating systems for de-icing, frost prevention and similar applications, and covers both indoor and outdoor systems.
		Examples of heating systems covered by 753 are heating systems for walls, ceilings, floors, roofs, drainpipes, gutters, pipes, stairs, roadways, and non-hardened compacted areas (e.g. football fields, lawns).
		The risks associated with ceiling heating systems are generally those of penetration of the heating element by nails, drawing pins, etc. pushed through the ceiling surface. For this reason, 753 requires that RCDs with a maximum rated residual operating current of 30 mA shall be used for automatic disconnection of supply (ADS). Similarly, there are concerns that under-floor heating installations can be damaged by carpet gripper nails, etc. For similar reasons, protection by a 30 mA RCD is required.
		To protect the building structure and provide precautions against fire, there are requirements to avoid overheating of the floor or ceiling heating system.
		In floor areas where contact with skin or footwear is possible, the surface temperature of the floor must be limited (for example, to 35 °C).
		Regulation 753.411.3.2 requires heating units manufactured without exposed-conductive-parts to be provided on site with a grid with spacing of not more than 30 mm, or other suitable conductive covering above the floor heating or below the ceiling heating and connected to the protective conductor of the installation.

Tables of symbols   B

▼ **Table B.1** General symbols

V	volts
A	ampere
I	current
Ω	ohm
Hz	hertz
W	watts
kW	kilowatts
C	capacitance
F	farads
uF	microfarad
p.u.	per unit
ph	phase
p.f.	power factor
L	line
N	neutral
h	hour
min	minutes
s	second
====	direct current (DC)
∼	alternating current (AC)

▼ **Table B.1** *cont.*

Symbol	Description
2 ∿	two-phase alternating current
2N ∿	two-phase alternating current with neutral
3 ∿	three-phase alternating current
3N ∿	three-phase alternating current with neutral
IPXX	IP number
⚡	fault (indication of assumed fault location)
▣	Class II appliance. Equipment with this mark indicates that additional measures have been taken to prevent the risk of the user receiving an electric shock. Commonly, this consists of an additional layer of insulated material protecting live parts (see BS 2754).
◇III◇	Class III appliance (equipment in which protection against electric shock relies on supply at separated extra-low voltage (SELV) and in which voltages higher than those of SELV are not generated) (see BS 2754).
(symbol)	Safety isolating transformer. Class III equipment must be supplied from a safety isolating transformer to BS EN 61558-2-6. The safety isolating transformer will have this identifying mark upon it.
(symbol)	Isolating transformer
⏚	Protective earth, general symbol (preferred to ⏚)

B

▼ **Table B.2** International system of units: base units

In the International System of Units (known as SI), there are seven base units, as shown below.

Quantity	Unit	Name of base unit	Unit symbol
Length	L	metre	m
Mass	m	kilogram	kg
Time	t	second	s
Electric current	I	ampere	A
Thermodynamic temperature	degree	kelvin	K
Amount of substance	mole	mole	mol
Luminous intensity	Cd	candela	I_V

▼ **Table B.3** Multiples and sub-multiples of quantities[1]

10^{18}	exa	E				10^{-3}	milli	m
10^{15}	peta	P	10^2	hecto	h	10^{-6}	micro	μ
10^{12}	tera	T	10^1	deca	da	10^{-9}	nano	n
10^9	giga	G	10^{-1}	deci	d	10^{-12}	pico	p
10^6	mega	M	10^{-2}	centi	c	10^{-15}	femto	f
10^3	kilo	k				10^{-18}	atto	a

[1] Powers in steps of 3 are preferred, but some others have common usage (e.g. centimetre: cm; decibel: dB).

▼ **Table B.4** SI derived units

The units of all physical quantities are derived from the seven base units and supplementary SI units and certain of them have been named. These, together with some common compound units, are given here:

Quantity		Unit name	Unit symbol	SI units
Force	F	newton	N	kg m/s²
Energy	E	joule	J	N m
Power	P	watt	W	J/s
Pressure, stress	p	pascal	Pa	N/m²
Electric potential	V	volt	V	J/C, W/A
Electric charge, electric flux	Q	coulomb	C	A s
Magnetic flux	ΦB	weber	Wb	V s
Magnetic flux density	B	tesla	T	Wb/m²
Resistance	R	ohm	Ω	V/A
Conductance	G	siemens	S	A/V
Capacitance	C	farad	F	C/V
Inductance	L	henry	H	Wb/A
Celsius temperature	c	degree Celsius	°C	K
Frequency	f	hertz	Hz	s⁻¹
Luminous flux	Φ_V	lumen	lm	cd sr
Illuminance	E_V	lux	lx	lm/m²
Mass density	m	kilogram per cubic metre		kg/m³
Torque	T	newton metre		N m
Electric field strength	E	volt per metre		V/m
Magnetic field strength	H	ampere per metre		A/m
Thermal conductivity	K	watt per metre kelvin		W m⁻¹K⁻¹
Luminance	sb	candela per square metre		cd/m²

▼ Table B.5 Symbols for use in schematic wiring diagrams

Symbol	Description
⊗	Transformer, general symbol
─•─	General wiring, with joint
─///─ Phase(s) ─•─ N ─┬─ Protective conductor ─┬─ Combined protective and neutral conductor (PEN)	identification of specific conductors
─[]─	Operating device, general symbol (coil)
─\─	Make contact, normally open
─/─	Break contact, normally closed
├--\─	Manually operated switch, general symbol
△	Three-phase winding, delta
─▷├─	Semiconductor diode, general symbol
Y	Three-phase winding, star

▼ **Table B.5** *cont.*

Symbol	Description
◿	Changer, general symbol Convertor, general symbol **NOTES:** 1 If the direction of change is not obvious, it may be indicated by an arrowhead on the outline of the symbol. 2 A symbol or legend indicating the input or output quantity, waveform, etc., may be inserted in each half of the general symbol to show the nature of the change.
∼/=	Rectifier
=/∼	Inverter
⊣⊦	Battery of primary or secondary cells
▯	Fuse link, rated current in amps, general symbol
⊣⊦	Capacitor, general symbol
‿‿‿‿	Inductor coil, winding or choke
‿‿‿‿	Inductor coil, winding or choke with magnetic core

B

▼ **Table B.6** Making and breaking current

	Switch
	Switch-fuse
	Fuse-switch

▼ **Table B.7** Isolating

	Isolator (disconnector), general symbol
	Disconnector-fuse (fuse combination unit)
	Fuse-disconnector
	Circuit-breaker suitable for isolation

▼ **Table B.8** Making, breaking and isolating

	Switch-disconnector
	Switch-disconnector-fuse (fuse combination unit)
	Fuse-switch-disconnector

B

▼ **Table B.9** Meters

Symbol	Description
(V)	Voltmeter
(A)	Ammeter
*	Integrating instrument or energy meter * Function Wh = watt-hour VArh = volt ampere reactive hour

▼ **Table B.10** Location symbols for installations

Symbol	Description
⊛	Machine, general symbol * Function M = Motor G = Generator
▽	Motor starter, general symbol
▽△	Star-delta starter
⊥	Socket-outlet, general symbol
⊥/	Switched socket-outlet
○—	Switch, general symbol
○/	2-way switch, single-pole
⋈	Intermediate switch

▼ Table B.10 cont.

Symbol	Description
	Pull switch, single-pole
	Lighting outlet position, general symbol
	Fluorescent luminaire, general symbol
	Emergency lighting luminaire (or special circuit)
	Self-contained emergency lighting luminaire
	Push-button with indicator lamp
	Lamp or signal lamp, general symbol
	Clock, general symbol
	Acoustic signalling device, general symbol (e.g. bell)
	Buzzer
	Telephone handset, general symbol
	Microphone
	Loudspeaker
	Antenna, general symbol
G	Static generator

Degrees of protection provided by enclosures (IP code)   C

The requirements of the IP code are given in BS EN 60529: 1992+A2:2013. For more information see IET Guidance Note 1 *Selection and Erection*.

The degree of protection provided by an enclosure is indicated by two numerals, followed by an optional additional letter and/or optional supplementary letter(s) as shown in Figure C.1.

▼ **Figure C.1** IP code format

```
                           IP   2   3   C   H
```

Code letters
(international protection)

First characteristic numeral
(numerals 0 to 6, or letter X)

Second characteristic numeral
(numerals 0 to 8, or letter X)

Additional letter (optional)
(letters A, B, C, D)

Supplementary letter (optional)
(letters H, M, S, W)

C

For the purposes of this Guide, IP codes cited are defined as follows:

IP2X Penetration by a solid foreign object ≥12.5 mm in diameter shall not be possible.

IPXXB Access of a finger shall not be possible.

IP2XC Penetration by a solid foreign object ≥12.5 mm in diameter shall not be possible. Additionally, an inserted 2.5 mm^2 probe of 100 mm in length shall have adequate clearance from live parts.

IP4X Penetration by a solid foreign object ≥1.0 mm in diameter shall not be possible.

IPXXD Access by a 100 mm length of wire with a cross-sectional area (csa) of 1.0 mm^2 shall not be possible.

IPX4 Water splashed against the enclosure from any direction will not affect the equipment.

IPX5 Water jets directed against the enclosure from any direction will not affect the equipment.

IPX7 Temporarily immersed enclosure, ingress of water shall not cause harmful effects to the equipment.

Answers

Section 1   Regulation numbering

Regulation 462.1.201

Part: 4

Chapter: 46 Isolation and switching

Section: 461 Isolation

Regulation: 462.1.201

Section 1   The IET Wiring Regulations

1. They were first issued in 1882.
2. The international committee responsible for electrotechnical regulations is the International Electrotechnical Committee (IEC).
3. 'Second fix' is one of the final stages of an electrical installation carried out near the end of construction.
4. Appendix 4 contains information about the various types of installation methods.
5. The UK committee responsible for the IET Wiring Regulations is the Joint Power Electrotechnical committee (JPEL/64).
6. Part 7.
7. The Electricity Safety, Quality and Continuity Regulations (ESQCR) 2002, as amended.
8. Part 2 of the IET Wiring Regulations contains the Definitions.
9. Personal Protective Equipment at Work Regulations.
10. PUWER 1998 (Provision and Use of Work Equipment Regulations).

Section 2 Health and Safety

1 HSE GS 38.
2 The insulation on tools designed for electrical work is to protect the user from electric shock.
3 For the purposes of your examinations, the key should be kept in your pocket. If you don't have a pocket, you should find a way to keep the key with you at all times.
4 False: a risk assessment should be periodically reviewed throughout the entire duration of the work being carried out.
5 Wire strippers. You may encounter experienced electricians using various other tools, such as side cutters or pliers. However, these are more likely to damage the conductor and cause it to break when terminating.

Section 3 Generation and Transmission

1 Any three from nuclear, solar, wind, hydro or wave.
2 Electricity Safety, Quality and Continuity Regulations (ESQCR) 2002, as amended.
3 Electricity (alternating current) can be transmitted much more efficiently when the voltage is increased.
4 Distribution transformers are generally found close to areas where the electricity is being used.
5 Electricity in the UK should be at 50 Hz.
6 Electricity is usually charged per kilowatt hour (kWh).
7 There are many aspects of the National Grid, but one of the greatest, mentioned in Section 3, is that the entire UK's power supply is connected together, so that if, for some reason, one power station fails, the towns and cities in that area will still have power.
8 Step-up transformers increase voltage.
9 Step-down transformers reduce voltage.
10 Maximum 253 V (230 + 10 %) and the minimum 216 V (230 - 6 %).

Section 4   Supply

1   Star connection and delta connection.
2   Protective earthed-neutral (a combined protective and neutral conductor in the cable supplying an installation).
3   Information must include:
    (a) nominal voltage;
    (b) nature of current and frequency;
    (c) prospective short-circuit current;
    (d) external EFLI;
    (e) suitability for maximum demand; and
    (f) type and rating of overcurrent device at the origin.
4   Single-core, insulated and sheathed cable.
5   Regulation 421.1.201.
6   Any suitable example given in Regulation 314.1.
7   Eight.
8   Regulation 536.4.5.
9   20 A.

Section 5   Protection and Isolation

1   An overcurrent occurring in a circuit that is electrically sound.
2   Asbestos.
3   The serviceable short-circuit current rating.
4   The approximate disconnection times are:
    (a) 1,000 seconds or 17 minutes;
    (b) 500 seconds or 8.5 minutes;
    (c) 100 seconds or 1.5 minutes; and
    (d) 250 seconds or 4 minutes.
5   30 mA.
6   Chapter 53.
7   Regulation 537.2.5: "Provision shall be made for securing off-load isolating devices against unwanted or unintentional opening. This may be achieved, for example, by locating the device in a lockable space or lockable enclosure or by padlocking. Alternatively, the off-load device may be interlocked with a load-breaking one."
8   Examples may include:

(a) light switches;
(b) time clocks;
(c) PIR sensors; and
(d) contactors.
9 The machine could suddenly re-start, should the power come back on following a power cut.
10 A contactor or direct-on-line starter unit having start and stop buttons (see 5.11 for details).

Section 6   Earthing and Bonding

1 No, the earthing conductor is not a live conductor.
2 Terre terre (TT) systems are usually the first choice for providing a means of earthing if the electricity supplier cannot provide a means of earthing.
3 In order for a protective device, such as a circuit-breaker, to operate, the fault current must have a good path to earth returning to the neutral conductor at the distribution transformer, so that it completes the earth fault path.
4 Earth electrode.
5 As below:
   (a) Terre Neutral-Separated: the earth is a separate conductor connected to the neutral terminal at the distribution transformer;
   (b) Terre Neutral-Connected-Separated: the earth and neutral conductors come into the property as one conductor and are separated at the consumer's installation; and
   (c) Terre Terre: this is where the general mass of the Earth is used to provide a path for the fault current to travel back to the neutral at the distribution transformer.
6 Exposed-conductive-part: this is part of a piece of electrical equipment or installation.
7 Extraneous-conductive-part: this is not part of the electrical installation, but it is in contact with the general mass of Earth and may provide a path for fault current to travel back to the neutral at the distribution transformer.
8 During a fault to earth.

9 Close to zero.

10 It is required so that if one of the connections to an extraneous-conductive-part becomes disconnected or damaged, the conductor will still provide a path for current to flow from the other extraneous-conductive-part(s) to the main earthing terminal of the installation.

Section 7   Cable Calculations

Task one

$I_b = P/V$, so $I_b = 9,000/230$

$I_b = 39.13$ A

Select I_n:

$I_n \geq 39.13$ (I_n = BS EN 60898 40 A (the next available size up from 39.13))

So $I_b \leq I_n$

Rating factors

$C_a = 0.94$

$C_i = 1$ can be used (by using reference method 100 from Table F6 of the *On-Site Guide* and/or Table 4D5 of the IET Wiring Regulations)

$C_f = 1$ (as the device being used is not a semi-enclosed fuse)

C_c = not applicable

C_d = not applicable

C_g = not applicable

C_s = not applicable

Select the right size cable (using Table 4D5 of the IET Wiring Regulations)

$I_t \geq \dfrac{I_n}{C_a \times C_i \times C_f}$

$I_t \geq \dfrac{40}{0.94 \times 1 \times 1}$

$I_t = 42.55$ A

According to Table 4D5 of the IET Wiring Regulations, a 10 mm² cable will be adequate to carry the design current when installed using reference method 100.

Check that the voltage drop is within the permitted values given in Table 4Ab of the IET Wiring Regulations.

$$Voltage\ drop = \frac{(mV/A/m \times I_b \times L)}{1{,}000}$$

$$Voltage\ drop = \frac{(4.4 \times 39.13 \times 6.3)}{1{,}000}$$

Volt drop = 1.08 V

As this is within the permissible value for volt drop (5 % for other uses), this suggests that the cable is suitable for the job.

Task two

$I_b = P/V$, so $I_b = 500/230$

$I_b = 2.17$ A

Select I_n

$I_n \geq I_b$

I_n = BS EN 60898 6 A (the next available size up from 2.17)

$I_n \geq 2.17$ A

Rating factors

$C_a = 1$

C_i = not applicable

$C_f = 1$ (as the device being used is not a semi-enclosed fuse)

C_c = not applicable

$C_d = 1.03$

C_g = not applicable

$C_s = 1$

Select the right size cable (Table 4D5 of the IET Wiring Regulations)

$$I_t \geq \frac{I_n}{C_a \times C_f \times C_d \times C_s}$$

$$I_t \geq \frac{6}{1 \times 1 \times 1.03 \times 1}$$

$I_t = 5.82$ A

According to Table 4D4A of the IET Wiring Regulations, a 1.5 mm² cable will be adequate to carry the design current when installed using reference method D.

Check that the voltage drop is within the permitted values given in Table 4D5 of the IET Wiring Regulations.

$$\text{Voltage drop} = \frac{(mV/A/m \times I_b \times L)}{1,000}$$

$$\text{Voltage drop} = \frac{(29 \times 2.17 \times 20)}{1,000}$$

Volt drop = 1.26 V

As this is within the permissible value for volt drop (5 % for other uses), this suggests that the cable is suitable for the job.

Section 8   Final Circuits

1. Fused and unfused.
2. 30 A or 32 A.
3. A fused connection unit with a 13 A fuse.
4. 250 volts: Regulation 559.5.1.201 states: "A ceiling rose or lampholder shall not be installed in any circuit operating at a voltage normally exceeding 250 volts".
5. Regulation 559.9.
6. As below:
   (a) off;
   (b) on;
   (c) off; and
   (d) on.

7 Extra-low voltage to reduce the severity of electric shock and separation of the supply to the equipment and the incoming electrical supply and earth.
8 PELV does not separate the circuit equipment from earth.
9 Part 7, Section 701.
10 In control systems.

Section 9   Inspection and Testing

1 An insulation resistance test will allow the tester to identify whether or not the insulation between the live conductors and live conductors and earth is sufficient.
2 The highest value is recorded.
3 The lowest value is recorded.
4 The line and earth (cpc) must be connected together (linked) at the consumer unit.
5 Maximum Z_s values can be found in Appendix B of the *On-Site Guide*.
6 The switches may need to be operated in order to obtain a reading.
7 The recommended length is 2 mm and they should be no more than 4 mm.

Section 10   Fault Finding

1 If a fault cannot be rectified, the circuit must be left in a safe condition until the fault can be rectified. For example, if additional material is required from the wholesaler, the electrician should make the installation safe before leaving.
2 A low resistance would indicate a satisfactory earth fault path.
3 The first logical step would be to split the circuit into two and retest each part of the circuit so that the part with the fault can be found (don't get confused between identifying and locating the fault).
4 Once a fault has been rectified, the sequence of tests must be carried out as they would on a new installation for any part of the installation that has been affected by the fault-finding process.

Section 11 Common Calculations

1. As below:
   (a) 12
   (b) 9
   (c) 3
2. Watts
3. I × R = V
4. 2 A
5. As below:
   (a) 9.5 Ω
   (b) 51 Ω
   (c) 15 Ω
6. (a) 2.1 Ω
   (b) 0.62 Ω

Section 12 Diversity

1. The protective device should be rated at 15.5 A or the next available higher rating, for example, a 16 A circuit-breaker to BS EN 60898.
2. As below:
   (a) The installer may over-engineer the installation at additional cost.
   (b) The customer would have an expensive installation that will never be used to its full capacity.
3. 100 % full load (f.l.) of the largest appliance + 80 % f.l. of the second largest appliance + 60 % f.l. of remaining appliances.
4. Any electrical appliance that is designed to run continuously at full load, such as a shower.
5. Guidance provided in the *On-Site Guide* asks installers to assume that there will be a demand of 100 W per lampholder on a lighting circuit.
6. The working should look something like this:

$I_n \geq I_b$

$I_b = \dfrac{P}{V}$

The total, P, is broken into the individual parts of the cooking appliance:

$P = (2 \times 2{,}000) + (2 \times 750) + 3{,}000$

$I_b = \dfrac{8{,}500}{230}$

$I_b = 37$ A

As stated in the question, this cooking appliance has a socket-outlet included, so an additional 5 A is added to I_b.

$37 + 5 = 42$

$I_b = 42$ A

Now that the total I_b has been worked out, we can apply diversity.

10 A is taken from I_b

$42 - 10 = 32$

$I_b = 32$ A

Only 30 % of the remaining 32 A is taken.

$32 \times 30\ \% = 9.6$ A

$I_b = 9.6$ A

At this stage, you can add the 5 A for the socket-outlet and the 10 A that was deducted at the beginning:

$I_b = 9.6 + 5 + 10$

$I_b = 24.6$ A

$I_n \geq 24.6$ A

The circuit-breaker to be installed for this circuit should have a rating of 30 A or 32 A.

Section 13 Prosumer's Electrical Installations

1. Chapter 82 contains the information and requirements for prosumer's installations.
2. Supply, distribution, consumption and energy management.
3. A generator and a load.
4.
   1. Direct feeding mode.
   2. Reverse feeding mode.
   3. Island mode.
5.
   1. Supply current-using equipment.
   2. Be charged by local power supplies.
   3. Be charged by the public network.
6.
   1. Supply current-using equipment.
   2. Supply local storage units.
   3. Supply the public network.
7. When operating in island mode.
8. Electrical energy management system.

Index

A

abbreviations	Table 1.3; Table 1.4
acceptable risk	2.2.4; Table 2.2
additions and alterations to an installation	8.2.1
AFDDs (arc fault detection devices)	9.5.7
ambient temperature	7.2.2; 7.3.2
appliance demands	12.3
arc fault detection devices (AFDDs)	9.5.7
awarding bodies	1.5.2

B

bonding	Section 6
breaking capacity	5.1
British Standards	Table 1.1
buried cables	7.2.2

C

cable ladders	Table 1.5
cables	
calculations	Section 7
current-carrying capacity	1.7.2
grouping	7.2.2
installing	8.2.3
lighting circuits	8.3.2
rating	1.7.3; 7.2; 7.3
routes	8.2.4
terminology	Table 1.5
types	1.5.1; Table 1.2
voltage drop	7.4
cable trays	Table 1.5
cartridge fuses	5.1.2
ceiling roses	8.3.2
CENELEC (European Committee for Electrotechnical Standardization)	Table 1.1
circuit-breakers	5.1.5
standards	Table 5.1
circuit design: *see final circuits; see lighting circuits*	
circuit protective conductors (cpc)	6.1.1
continuity testing	9.4.2; 9.4.3
ring final circuits	8.5.1; 8.5.2
circuit protective devices: *see protective devices*	
competent person scheme	Table 1.3

Index

conduits	Table 1.5
consumer's tails	4.6
consumer units	4.7
heights	8.2.3
installing	8.2.1; 8.2.2
ways and modules	4.7.1
continuity testing	9.4.1; 9.4.2
cpc: see *circuit protective conductors (cpc)*	
cross polarity	10.1.3; 10.2.5
cross-sectional areas (csas)	8.5.2; 8.5.4
current-carrying capacity	1.7.2
current causing effective disconnection	5.1
current rating	
of cables and equipment	1.7.2; 7.2; 7.3
of supply	4.3

D

dead tests	9.4
design current	7.2.1; 7.3.1; 7.3.2
disconnection current	5.1
discrimination: see *selectivity*	
distribution board (DB)	4.7
distribution circuits	4.7
distribution network operators (DNOs)	4.3.1; 6.1
distribution networks	3.3.3
distribution system operators (DSOs)	4.3.2
distribution transformers	3.3.3
diversity	Section 12

DNOs (distribution network operators)	4.3.1; 6.1
DSOs (distribution system operators)	4.3.2

E

earth electrodes	4.4.1; 4.4.3; 6.1.4
earth fault loop	6.2.4
earth fault loop impedance (EFLI)	1.7.4; 8.2
testing	9.5.2; 9.5.3
earth faults	5.1; 6.1.2; 6.2.3; 10.2.2
earthing	Section 6
arrangements	4.4; 6.1.4
protective equipotential bonding	6.2
responsibility for	6.1
supplementary equipotential bonding	6.3
EFLI: see *earth fault loop impedance (EFLI)*	
electrical isolation: see *isolation*	
Electricity at Work Regulations 1989 (EAWR)	9.1.1
electricity generation	Section 3
electricity meters	3.4
Electricity Safety, Quality and Continuity Regulations (ESQCR)	1.5.3
electricity transmission	3.3
electric shock	6.2.3; 6.2.4
emergency switching	5.9
environmental conditions	8.3.3

Index

equipotential bonding: *see protective equipotential bonding conductors*	
escape routes	8.2.4
ESQCR (Electricity Safety, Quality and Continuity Regulations)	1.5.3
European Committee for Electrotechnical Standardization (CENELEC)	Table 1.1
European Standards	1.2; Table 1.1
exposed-conductive-part	Table 1.2
extra-low voltage	1.5.1; 8.8; 8.9; 8.10; Table 1.2
extraneous-conductive-part	Table 1.2

F

fault finding	Section 10
FELV (functional extra-low voltage systems)	8.10
final circuits: *(see also radial final circuits); (see also ring final circuits)*	4.7; Section 8
additions and alterations to an installation	8.2.1
cable routes	8.2.4
consumer unit arrangements	8.2.2
designing and installing new circuits	8.2
functional extra-low voltage systems (FELV)	8.10
general installation practices	8.1
heights of switches, socket-outlets and consumer units	8.2.3
luminaires and lighting installations	8.3
protective extra-low voltage systems (PELV)	8.9
separated extra-low voltage systems (SELV)	8.8
shower circuits	8.7
spurs	8.6
Fire safety	8.2.4
functional extra-low voltage systems (FELV)	8.10
functional switching	5.10
fuse board	4.7
fuse carrier	4.3.1
fuses	
and cable rating	7.2.2
cartridge	5.1.2
high rupturing capacity (HRC)	5.1.1
plug-top	5.1.3
semi-enclosed	5.1.4
service	4.3.1
standards	Table 5.1

G

generators	Section 3
Guidance Note GS38	2.3.1

Index

H

hand tools	2.4
	Table 2.3
health and safety: *(see also electric shock)*	Section 2
Health and Safety at Work etc. Act 1974 (HSWA)	2.1
Health and Safety Executive	1.5.3; Table 1.4
heights of fittings	8.2.3
high resistance	10.1.2; 10.1.5; 10.2.4
high rupturing capacity (HRC) fuses	5.1.1
hydro generation	3.1.2

I

I_{CN} rating	5.1.5
I_{CS} rating	5.1.5
IEC (International Electrotechnical Commission)	Table 1.1
IET Wiring Regulations	1.1
In Home Display (IHD)	3.4
initial inspection	9.2; Table 9.1
inspection: *(see also testing)*	Section 9
initial inspection	9.1.3; 9.2
legal aspects	9.1.1
responsibilities	9.1.2
installation	1.7.1; 8.1; 8.2
location symbols	Table B.10
institutions	1.5.2; Table 1.3
insulation resistance	
minimum values	9.4.4; Table 9.2
testing	9.4.4
integration of devices and components	5.4
interconnection of conductors	10.1.4; 10.2.6
intermediate light switches	8.3.2
International Electrotechnical Commission (IEC)	Table 1.1
IP code	8.3.2; Appx C
isolating transformers	8.8.1; 8.9.1
isolation: *(see also switching)*	5.5
for mechanical maintenance	5.8
for particular installation types	5.7; Table 5.3
symbols	Table B.7; Table B.8
isolators	
compared with switches	5.6
isolation switch (mains)	4.5; 4.7

J

JPEL/64 Committee	1.3

K

kitchen appliance demands	12.3

L

legal aspects	1.6; 9.1.1
lighting circuits	8.3.2; 9.4.2

Index

lighting installations	8.3
environmental conditions	8.3.3
regulations and guidance	8.3.1
selection of equipment	8.3.2; 8.3.3
light switches	8.3.2
link method	9.4.2
live conductor arrangements	4.2; Table 4.1
live tests	9.5
long lead method	9.4.2
luminaires	8.3; 8.3.2

M

main switch	4.7
maintenance, switching off for	5.8
materials	Table 1.5
maximum demand: *(see also diversity)*	4.3
mechanical maintenance	5.8
meters	3.4; Table B.9
meter tails	4.6
method statements	2.2.3
miniature circuit-breakers (MCBs)	5.1.5
motor circuits	5.11
moulded case circuit-breakers (MCCBs)	5.1.5
multiples and sub-multiples of quantities	Table B.3

N

National Grid	3.3
Neutral-to-earth faults	5.1; 8.2.1
nominal rating	5.1

O

Ohm's law	11.2
open circuits	10.2.3
organizations	1.5.2; Table 1.3
overcurrent	5.1
overload current	5.1; 10.1.6
overload protection	7.2.2

P

parallel circuits	11.3.2
PEIs (prosumer's electrical installations)	13
PELV (protective extra-low voltage systems)	8.9
PEN (protective earthed-neutral) conductor	4.4.1; 6.1.4
Personal Protective Equipment at Work Regulations 1992 (PPEWR)	Table 1.4
phase sequence testing	9.5.5
plug-top fuses	5.1.3
PME (protective multiple earth)	4.4.1; 6.1.4
polarity testing	9.4.5; 9.5.1
policies	1.6
potential difference	6.2.2
power distribution	3.3.3
power generation	Section 3
power law	11.2
power sources	3.1
power stations	3.1.1
power supply characteristics	3.2

Index

power tools	2.4; Table 2.4
power transmission	3.3
PPEWR (Personal Protective Equipment at Work Regulations 1992)	Table 1.4
pre-determined values	1.7
prospective fault currents (PFC)	5.1; 5.1.5; 5.3
testing	9.5.4
prosumer's electrical installations (PEIs)	Section 13
protected escape routes	8.2.4
protective conductor current	5.1
protective conductors: *see circuit protective conductors (cpc)*	
protective devices: *(see also circuit-breakers); (see also fuses)*	4.7; 5.1
current rating	5.1; 7.2; 7.3
selecting	5.2
standards	Table 5.1; Table 5.2
protective earthed-neutral (PEN) conductor	4.4.1; 6.1.4
protective equipotential bonding conductors	6.2
continuity testing	9.4.1
protective extra-low voltage systems (PELV)	8.9
protective multiple earth (PME)	4.4.1; 6.1.4

Q

qualifications	1.3.2

R

radial final circuits	8.4
continuity testing	9.4.2
maximum area served	8.4.5
maximum number of socket-outlets	8.4.3
spurs	8.6
rated short-circuit capacity	5.1
rating factors	1.7.3; 7.2.2
Residual-current circuit-breaker (with overcurrent protection) (RCBO)	5.1.7
residual current devices (RCDs)	5.1.6; 8.2.2
testing	9.5.6
resistors	11.3
rewirable fuses: *see semi-enclosed fuses*	
ring final circuits	8.5
circuit protective conductors (cpc)	8.5.1
continuity testing	9.4.3
diversity	12.3
fault finding	10.2.7
maximum area served	8.5.4
maximum number of socket-outlets	8.5.3
spurs	8.6
risk assessments	2.2; 9.1.2

Index

S

safe isolation procedure	2.3	protective devices	Table 5.1; Table 5.2
selectivity	5.3	statutory requirements	1.6
semi-enclosed fuses	5.1.4; 7.2.2	substations	3.3.2
separated extra-low voltage systems (SELV)	8.8	supplementary equipotential bonding	6.3
series circuits	11.3.1	supply intake	4.1; 4.5
service fuses	4.3.1	current rating	4.3
short circuits	5.1; 10.1.1; 10.2.2	earthing arrangements	4.4
shower circuits	8.7	live conductor arrangements	4.2; Table 4.1
single-phase supply	4.2	responsibility	4.5
SI units	1.4; Table B.2; Table B.4	tails	4.6
smart meters	3.4	supply rating	4.3
socket-outlets: *(see also radial final circuits)*; *(see also ring final circuits)*		surge protection devices (SPDs)	9.5.7
		switches	
		compared with isolators	5.6
heights	8.2.3; 8.4.4	heights	8.2.3; 8.4.4
maximum number	8.4.3; 8.5.3	lighting circuits	8.3.2
spurs	8.6	symbols	Table B.6; Table B.8
soil resistivity	7.2.2		
solar power	3.1.3	switching: *(see also isolation)*	5.5
SPDs (surge protective devices)	9.5.7	emergency	5.9
		functional	5.10
special installations/ locations	Appx A	undervoltage protection	5.11
SPDs (surge protection devices)	9.5.7	symbols	1.4; Appx B
split-way boards	4.7.1	**T**	
spurs	8.6	tails	4.6
standards	Table 1.1	terminology	1.5.1; Table 1.5; Table 1.6
power supply characteristics	3.2	testing	Section 9

Index

dead tests	9.4
initial inspection	9.1.3; 9.2
legal aspects	9.1.1
live tests	9.5
responsibilities	9.1.2
thermal insulation	7.2.2
three-phase supply	4.2
three-plate method	8.3.2
TN-C-S systems	4.4.1; 6.1; 6.1.4
TN-S systems	4.4.2; 6.1.4
tool safety	2.4
touch voltage	6.2.3; 6.2.4
transmission networks	3.3.2
transposition	11.1
triangle method	11.2
trunking	Table 1.5
TT system	4.4.3; 4.7
TT systems	5.1.6; 6.1.1; 6.1.4
two-plate method	8.3.2
two-way light switches	8.3.2

U

uninterruptible power supply (UPS)	13.1
undervoltage protection	5.11

V

verification of installation	9.1.3
visual inspection	10.2.1
voltage drop	7.4

W

wander lead method	9.4.2
wind generation	3.1.2
wiring diagrams	Table B.5

Have you found this book useful?
Find more guidance like this from the IET

The IET publishes a range of guidance for use by the experienced engineer and those new to the electrical industry.

Visit our webshop to find more, including:

- Electrician's Guide series
- Student's Guides
- Guidance Notes to BS 7671

As well as a range of Codes of Practice and dedicated guidance on other areas of electrical installation.

See **theiet.org/regs-books** for our complete range

To find out about subscriptions to access our content online for you or your company, visit **theiet.org/wrd-mp**

The Institution of Engineering and Technology (IET) is registered as a Charity in England and Wales (No. 211014) and Scotland (No. SC038698). The Institution of Engineering and Technology, Michael Faraday House, Six Hills Way, Stevenage, Hertfordshire SG1 2AY, United Kingdom.

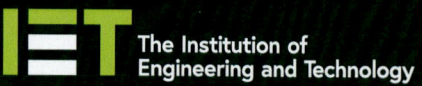

Academy

Complement your learning from our electrical guides with online courses from the IET Academy

The IET Academy provides engaging, interactive e-learning, covering a range of key electrical industry topics, including BS 7671, Surge Protection and In-service Inspection and Testing.

E-learning gives you the opportunity to study at your own pace, in your own time. Flexible mobile access allows you to take a course in full or in shorter bite-size units, and course content is interactive and in-depth, providing the same level of learning you'd get in a classroom.

Our courses are designed to provide learning that is relevant to what you do. Courses are available in the following areas:

ELECTRICAL	COMMUNICATIONS	POWER	PROFESSIONAL SKILLS	SAFETY & SECURITY	TRANSPORT	WIRING REGULATIONS

Find out more at theiet.org/academy-mp

The Institution of Engineering and Technology (IET) is registered as a Charity in England and Wales (No. 211014) and Scotland (No. SC038698). The Institution of Engineering and Technology, Michael Faraday House, Six Hills Way, Stevenage, Hertfordshire SG1 2AY, United Kingdom.